JALE TOSUN AND
CHRISTIN HEINZ-FISCHER

EUROPEAN CLIMATE PACT AMBASSADORS

Motivations, Strategies, and Experiences

BRISTOL
UNIVERSITY
PRESS

First published in Great Britain in 2026 by

Bristol University Press
University of Bristol
1-9 Old Park Hill
Bristol
BS2 8BB
UK
t: +44 (0)117 374 6645
e: bup-info@bristol.ac.uk

Details of international sales and distribution partners are available at bristoluniversitypress.co.uk

© Jale Tosun and Christin Heinz-Fischer 2026

The digital PDF and ePub versions of this title are available open access and distributed under the terms of the Creative Commons Attribution-NonCommercial-NoDerivatives 4.0 International licence (https://creativecommons.org/licenses/by-nc-nd/4.0/) which permits reproduction and distribution for non-commercial use without further permission provided the original work is attributed.

DOI: 10.51952/9781529255850

British Library Cataloguing in Publication Data
A catalogue record for this book is available from the British Library

ISBN 978-1-5292-5583-6 paperback
ISBN 978-1-5292-5584-3 ePub
ISBN 978-1-5292-5585-0 OA PDF

The right of Jale Tosun and Christin Heinz-Fischer to be identified as authors of this work has been asserted by them in accordance with the Copyright, Designs and Patents Act 1988.

All rights reserved: no part of this publication may be reproduced, stored in a retrieval system, or transmitted in any form or by any means, electronic, mechanical, photocopying, recording, or otherwise without the prior permission of Bristol University Press.

Every reasonable effort has been made to obtain permission to reproduce copyrighted material. If, however, anyone knows of an oversight, please contact the publisher.

The statements and opinions contained within this publication are solely those of the authors and not of the University of Bristol or Bristol University Press. The University of Bristol and Bristol University Press disclaim responsibility for any injury to persons or property resulting from any material published in this publication.

Bristol University Press works to counter discrimination on grounds of gender, race, disability, age and sexuality.

Cover design: blu inc
Front cover image: iStock/designer491

Contents

List of Figures and Tables		iv
List of Abbreviations		v
Preface		vi
one	Introducing the European Climate Pact and its Ambassadors	1
two	Climate Action Through the Lens of the European Climate Pact	22
three	Conceptual Framework	47
four	Research Design and Operationalization	71
five	Empirical Findings	87
six	Conclusions and the Way Forward	132
Notes		150
References		153
Index		165

List of Figures and Tables

Figures

2.1	Areas in which the CPAs work	28
5.1	CPA population, October 2023	89
5.2	Survey respondents, November 2023	90
5.3	Motivational drivers for the decision to become a CPA	92
5.4	Motivational factors by professional background	97
5.5	Most important factor to become a CPA by gender, age group, and mandate	100
5.6	Absolute frequency of advocacy patterns	106
5.7	Different ways of pursuing climate action by mandate since the beginning of CPA term	109
5.8	Taking on climate action by occupational status	111
5.9	Access to political stakeholders by mandate	115
5.10	Access to political stakeholders by occupation	118
5.11	CPAs' experience with the mandate	121
5.12	Rating of experience by motivational drivers	123
5.13	Experience with mandate and internal as well as external political efficacy	126

Table

3.1	Conceptual framework	67

List of Abbreviations

CID	Clean Industrial Deal
CPA	Climate Pact Ambassador
ECI	European Citizens' Initiative
ECP	European Climate Pact
EGD	European Green Deal
EU	European Union
GHG	Greenhouse gas
MP	Member of Parliament
NGO	Non-Governmental Organization
UNFCCC	United Nations Framework Convention on Climate Change

Preface

Governance is fundamental to how climate and environmental policy is formulated and implemented within the European Union (EU). While this area has received extensive scholarly attention, the EU continues to introduce innovative governance designs that merit further investigation. The European Climate Pact (ECP) and its ambassador programme represent one such innovation, which particularly interests us because it recognizes the role of individual citizens in governance processes.

Having worked within the tradition of classic EU multilevel system governance, the creation of ECP ambassadors prompted us to shift our analytical focus from collective to individual actors. This intellectual journey enabled us to explore concepts drawn from political science, sociology, and community and social psychology. Now that we recognize the existence of governance formats that depend on individual actions, we advocate for future research on climate and environmental governance to emphasize the role of individuals within these arrangements and processes. We intend to pursue this line of academic inquiry in our ongoing research.

The analytical insights presented in this book would not have been possible without support from numerous individuals. We are particularly grateful to David Rzepka, who compiled the dataset that underpins this book's empirical analysis. As an undergraduate political science student at Heidelberg University, David demonstrated exceptional academic qualities when preparing and implementing the data collection. He has since progressed to the University of Göttingen for his graduate studies, and we are confident he will achieve considerable success in academia or any other field he chooses to pursue. We also extend our thanks to Leonardo Di Filippo, Team Leader of the ECP in the European Commission's Directorate-General for Climate Action, for his support of our data collection efforts. Additionally, we extend our gratitude to all Climate Pact Ambassadors who participated in our survey and whose dedicated efforts to engage with diverse communities are instrumental in delivering the European Green Deal.

The manuscript benefitted from feedback received during Dave Huitema's inauguration ceremony at the University of Wageningen and from participants in the Standing Group on Public Policy presentation of the European Consortium for Political Research, led by Karin Ingold, Johanna Hornung, Johanna Kuenzler, and Vilém Novotny. We also received constructive comments from Spyros Bakas, Laurence Crumbie, Colette Vogeler, and three anonymous reviewers, for which we are grateful.

Our collaborative research projects on climate and environmental governance have indirectly enriched this book, including: Deep Decarbonisation: The Democratic Challenge of Navigating Governance Traps (DeepDCarb; European Research Council; Grant Agreement Number: 882601), Accelerating Climate Action and the State: Getting to Net Zero (ACCELZ; Research Council of Norway; Grant Agreement Number: 335073), and Rethinking Co-creation of Digital and Environmental Policy in Systems of Multilevel Governance (RECODE MLG; European Commission; Grant Agreement ID: 101177521).

Our research and networking activities within the framework of the Jean Monnet Network of Excellence GreenDeal-Net (The European Green Deal: Governing the EU's Transition towards Climate Neutrality and Sustainability) have particularly influenced this research. We also gratefully acknowledge GreenDeal-Net's generous funding support for making this book permanently open access. The views and opinions expressed in this book are, however, ours alone and do not necessarily reflect those of the EU or the European Education and Culture Executive Agency (EACEA). Neither the EU nor EACEA can be held responsible for them.

Jale Tosun and Christin Heinz-Fischer
Heidelberg, July 2025

ONE

Introducing the European Climate Pact and its Ambassadors

Introduction

The European Union (EU) has emerged as one of the most proactive jurisdictions in adopting and implementing policies specifically designed to address environmental degradation and mitigate climate change (Cifuentes-Faura, 2022). It has demonstrated consistency and change in its commitment to climate and environmental action over the last decades (Knill and Liefferink, 2007; Zito et al, 2019; Lenschow et al, 2020; von Homeyer et al, 2022; Tosun, 2023; Bocquillon, 2024).[1] The EU's environmental and climate leadership extends beyond its own borders through a dual-pronged approach: the EU actively worked to harmonize and elevate the corresponding policies of its member states through internal governing mechanisms and has sought to influence and shape the policies of third countries through its external governance efforts (Knill and Tosun, 2009; Schulze and Tosun, 2013). These initiatives represent deliberate attempts to project the European policy approach globally and to exert meaningful leadership at the international level (Oberthür and Roche Kelly, 2008; Oberthür and Dupont, 2021; Tobin et al, 2023a).

The EU's policy approach to these issues was elevated to an unprecedented level of ambition when, on 11 December 2019, the European Commission, under the leadership of its President Ursula von der Leyen, published a landmark communication that formally introduced and outlined the

European Green Deal (EGD). This comprehensive policy framework represents a paradigm shift in EU governance, as the EGD functions as an ambitious economic growth strategy and a transformative climate and environmental agenda (Dupont et al, 2024). The overarching goal of the EGD is particularly ambitious in its scope: to 'transform the EU into a fair and prosperous society, with a modern, resource-efficient and competitive economy where there are no net emissions of greenhouse gases (GHGs) in 2050 and where economic growth is decoupled from resource use' (European Commission, 2019). This vision fundamentally reimagines the relationship between economic development and environmental sustainability, aiming to position the EU as a global pioneer in demonstrating that prosperity and environmental stewardship can be mutually reinforcing rather than competing objectives.

The European Commission explicitly indicates in the communication that the successful delivery and implementation of the EGD will necessitate the development and deployment of a comprehensive and interconnected set of policies designed to facilitate transformative change across multiple critical economic sectors (Dupont et al, 2024). These sectors encompass energy systems and renewable transitions, infrastructure development and modernization, transport networks and mobility solutions, food production and sustainable agriculture practices, as well as construction industries and building standards.

Additionally, the transformative approach of the EGD extends beyond sectoral policies to include fundamental reforms of taxation frameworks and social benefits systems, recognizing that achieving these ambitious objectives requires not only technological and industrial transformation but also corresponding adjustments to fiscal policies and social support mechanisms. This multisectoral approach reflects the European Commission's understanding that environmental and climate challenges are inherently cross-cutting issues that cannot be addressed through isolated policy interventions, but rather demand coordinated action across the spectrum of economic activity.

Public policies are not merely technical instruments or administrative tools but serve as fundamental expressions of how governments conceptualize, interpret, and frame certain societal problems while simultaneously revealing what they prioritize in their policy agendas and resource allocation decisions. Building on this analytical perspective, Boasson and Tatham (2023) present the compelling argument that climate policies inherently reflect the underlying purpose and strategic intent that policy makers attribute to them, which can be categorized into three distinct but potentially overlapping objectives: efficiency enhancement, systemic transformation, or public acceptance and legitimacy building.

When policy makers primarily aim to address market inefficiencies, particularly in situations where traditional market mechanisms fail to adequately account for environmental costs and negative externalities, they typically gravitate toward market-based policy instruments, such as emissions trading systems that seek to internalize these previously unaccounted costs. The EU's Emissions Trading System exemplifies this approach and remains the cornerstone of the EU's broader climate policy architecture, and continues to function as an important and enduring component of the EGD framework (Dupont et al, 2024).

Conversely, when policy makers seek to fundamentally reshape key economic sectors and stimulate comprehensive, long-term systemic change towards sustainable, low-carbon futures, they must deliberately choose more interventionist transformative policies. These include strategic industrial policy initiatives, targeted green job creation programmes, and sector-specific transition strategies that go beyond market corrections to actively restructure economic systems. Examples of policy mixes for low-carbon electricity transitions can be found in Germany and the UK, with the latter aiming for deeper transformation for certain periods (Geels et al, 2016).

Finally, when policy makers aim to build or maintain public support for climate and environmental policies, recognizing that policy success often depends on social acceptance, they need to strategically address the distributive effects and social implications of these policies while simultaneously creating legitimacy and public trust. This requires implementing comprehensive information campaigns, meaningful public participation processes, and carefully designed compensatory measures and mechanisms specifically aimed at facilitating a 'just transition' (Boasson and Tatham, 2023).

The driving force behind this third policy type, focused on acceptance and legitimacy building, is the increasingly pronounced politicization of climate policy in particular and the significant electoral backlash and popular resistance it has generated at the national level across various EU member states and at the supranational EU level itself (Patterson, 2023). This politicization has manifested itself in growing public scepticism towards climate policies perceived as economically burdensome or socially inequitable.

The most prominent and widely documented case of organized opposition to climate-related policy measures is undoubtedly the yellow vests (*gilets jaunes*) movement in France, which was initially sparked by widespread public opposition to a proposed increase in carbon tax on fuel. This grassroots movement quickly evolved into a broader expression of social and economic discontent that extended far beyond the original tax issue. The members and supporters of this movement articulated a fundamental sense of injustice, arguing that it was fundamentally unfair to expect them to bear the disproportionate financial burden of ambitious climate action policies, particularly given that their economic precarity and social vulnerability had already significantly deteriorated over the preceding decades due to various economic pressures, labour market changes, and austerity measures (Driscoll, 2023).

A more recent example is the wave of protests across several EU member states in late 2023 and early 2024 by farmers who opposed the EU's climate and environmental policies proposed or adopted under the EGD and called for easing their regulatory burden (Nagel et al, 2025).

Both cases illustrate the importance of addressing distributional concerns and ensuring that climate policies do not exacerbate existing social inequalities or impose undue hardships on already vulnerable populations, highlighting why acceptance-oriented policies have become increasingly essential for maintaining democratic legitimacy and public support for climate and environmental governance, especially given the growing strength of right-wing populist parties (Boasson and Tatham, 2023; Tatham and Peters, 2023).

Demonstrating awareness of the need to create and maintain democratic legitimacy and public support for the EGD, particularly in light of growing climate policy politicization across Europe, the European Commission has strategically built into the very core of its reform agenda a dedicated component that explicitly strives to achieve precisely this legitimacy-building objective. The component of particular interest and significance is the EGD's European Climate Pact (ECP), which was formally launched by the European Commission in December 2020 as a direct response to the recognition that top-down climate policies require bottom-up social acceptance to be effective and sustainable (European Commission, 2020b).

As the name suggests, the ECP focuses predominantly on climate action, although its scope extends beyond this core mandate. While climate action remains the primary emphasis, the ECP framework also addresses broader environmental challenges that are fundamentally intertwined with climate objectives, recognizing that these domains are interconnected but not synonymous. This integrated approach recognizes that effective climate action cannot be pursued in isolation from the wider environmental challenges facing Europe, which aligns with the overarching rationale of the EGD. The EGD has further strengthened the social dimension of climate and environmental action through the adoption of the Just Transition Mechanism and the Just Transition Fund, which aim to alleviate the socio-economic impact of the transition (Arabadjieva and Bogojević, 2024; Dupont et al, 2024; Sandmann et al, 2024). At the same time, the ECP's approach aligns with an observation made in the literature that climate change has come to dominate the EU's internal environmental policy agenda (Zito et al, 2019).

The ECP represents the EGD's primary institutional tool and mechanism for fostering the connection of citizens, communities, and organizations (Tosun, 2022c; Tosun et al, 2023a; Tosun et al, 2023b) with each other in order to exchange knowledge and develop solutions (Buzogány et al, 2025; Çelik, 2025). It cultivates citizen-led climate engagement and fosters authentic grassroots mobilization across the EU. This citizen engagement strategy is implemented through a network of dedicated Climate Pact Ambassadors (CPAs), who serve as voluntary representatives and change agents.

These CPAs are individual volunteers based throughout the 27 EU member states who formally commit to a threefold mission: to 'lead by example' through their own climate conscious behaviours and initiatives, to 'inspire others' by sharing knowledge and motivating action within their communities, and to 'connect stakeholders' by building bridges between different actors and facilitating collaborative climate action at the local and regional levels (European Commission, 2025b).

The roles assigned to CPAs are interesting to be reflected on from the perspective of 'green European citizenship' that recognizes the existence of private and public duties. According to Machin and Tan (2024), the ECP calls for changing lifestyle and consumption habits, and eating, travelling, and communicating differently, which are known as 'private duties'. However, it also calls for 'public duties' in the sense that citizens become involved in discussions that can stimulate technological and social innovation, improve decision making, as well as create co-ownership. The European Commission's conception of CPAs aligns with the private and public dimensions of citizenship.

This book specifically focuses on the ECP and particularly on the CPAs for several reasons that make these governance formats theoretically and practically relevant. First, the ECP represents a genuinely innovative governance approach within the broader EU institutional context and policy framework. While ambassador programmes specifically focusing on environmental and climate concerns have indeed existed for several years in various EU member states (see, for example, Søgaard Jørgensen and Pedersen, 2015), the ECP represents a significant scaling-up and institutionalization of this approach at the supranational level.

Beyond this scaling-up, the ECP also signals a broader institutional transformation in how the EU engages with civil society. In the past, the European Parliament was regarded as the key institution within the EU's political system that supported civil society (Crespy and Parks, 2019), for example, through its responsiveness to policy demands expressed through the European Citizens' Initiative (Tosun et al, 2022). The ECP represents a departure from this traditional model, as the new approach of supporting individuals as ambassadors introduces a more direct dimension to EU interactions with civil society.

Second, precisely because the ECP constitutes a new and potentially transformative governance tool, one that is fundamentally based on the principle of individuals voluntarily choosing to become CPAs without financial incentives or formal obligations, it gives rise to numerous scientifically relevant research questions that deserve systematic investigation. These include, as will be explored and discussed in greater detail later, the underlying motivational drivers and personal factors that lead individuals to volunteer for this particular climate engagement programme, the diverse strategies and approaches that CPAs develop and employ in order to effectively deliver on their multifaceted mandate, and their lived experiences, challenges, and reflections on participating in this novel form of climate governance.

Third, the ECP represents part of a broader and increasingly significant trend, because the European Commission is making more frequent and strategic use of 'pacts' as governance instruments for EU policy implementation and stakeholder engagement, as evidenced by initiatives such as the Rural Pact launched in 2021 (Parreira and Pires, 2025) and other similar collaborative frameworks that seek to bridge the gap between EU level policy ambitions and policy delivery on the ground.

European Climate Pact and Climate Pact Ambassadors

When the European Commission unveiled the EGD, it did more than simply outline ambitious policy targets. Recognizing that the transformation required to achieve climate neutrality by 2050 could not be accomplished through policy measures alone, the European Commission simultaneously announced the launch of the ECP (European Commission, 2020). This initiative represented a fundamental acknowledgement that the EU's climate ambitions would only be realized through the participation and commitment of citizens, communities, and organizations across the member states.

The ECP comprises three interconnected pillars of public engagement. The first pillar focuses on knowledge and awareness, aiming to encourage 'information sharing, inspiration, and foster public understanding of the threat and the challenge of climate change and environmental degradation and on how to counter it' (European Commission, 2019). This information-based and educational dimension recognizes that effective climate action requires citizens to understand the science behind climate change and the practical solutions available to address it.

The second pillar centres on creating institutional opportunities for meaningful participation. Rather than simply informing citizens about climate challenges, the European Commission committed itself to establishing concrete venues where people could actively contribute to climate solutions.

This represents a shift from top-down communication to collaborative engagement, acknowledging citizens as partners rather than passive recipients of climate policy.

The third pillar emphasizes capacity-building and empowerment, particularly targeting grassroots initiatives focused on climate action and environmental protection. This approach recognizes that sustainable change often emerges from local communities and that institutional support can amplify the impact of citizen-led initiatives.

The institutional design of the ECP deliberately builds on existing participatory mechanisms within the EU. As the European Commission (2019) outlined, the initiative would leverage 'on-going series of citizens' dialogues and citizens' assemblies across the EU, and the role of social dialogue committees'. This foundation provides legitimacy and practical experience in citizen engagement, ensuring that the ECP would not operate in isolation but rather complement and enhance existing democratic processes. Likewise, the ECP has become a participatory mechanism whose approach the European Commission has advocated for broader application in climate governance, as evidenced in key legislative documents, such as the European Climate Law (Oberthür and Kulovesi, 2025).

The ECP evolved into a multilayered platform offering various levels of citizen engagement, each designed to accommodate different capacities, interests, and commitments. At the foundational level, citizens can subscribe to the ECP newsletter, providing regular updates on climate developments and opportunities for engagement. The ECP also facilitates participation in the 'Monitor and cut your carbon footprint!' campaign, developed in partnership with the United Nations' ActNow initiative, which enables individuals to track their environmental impact through a mobile application.

For those seeking more active involvement, the ECP supports the organization of group climate activities, allowing citizens to mobilize their immediate networks around specific climate actions. The platform also provides curated resources to help participants spread clear and accurate information about climate change and available solutions, transforming engaged citizens into informed advocates within their communities. The ECP further enables the organization of satellite events, creating opportunities for localized climate action that connects to the broader European movement while remaining responsive to local contexts and priorities.

The two most intensive forms of ECP participation are the CPAs and the Partners of the Pact initiative, both requiring significant commitment. The CPAs, the focus of this book, embody the human face of the initiative, individuals who dedicate themselves to promoting climate action within their communities and professional networks. These ambassadors serve as bridges between EU climate and environmental policy and grassroots action, translating broad policy objectives into concrete, locally relevant initiatives. The ambassador programme will be presented and discussed in detail in Chapter 2.

The Partners of the Pact programme represents a more recent innovation, specifically targeting formal organizations rather than individuals. This programme emerged from the ECP Secretariat's experience with the ambassador format and reflects an institutional learning process that has shaped the ECP's evolution. Initially, CPAs could serve either in their personal capacity or as representatives of organizations, creating some ambiguity about institutional versus individual commitment (Tosun et al, 2023). The introduction of the Partners programme addresses this complexity by creating distinct formats for organizational and individual engagement.

Currently in its initial phase, the Partners programme involves a carefully selected group of organizations working closely with the ECP Secretariat to develop best practices and operational frameworks. This collaborative approach to programme development means that detailed information about Partners remains limited in the public domain, as the Secretariat prioritizes learning and refinement over immediate scale.

Another form of participation in the ECP is the format 'Friends of the Pact', which provides an entry point for people who want to support and promote the ECP and its goals without committing to the more structured role of being an official CPA. Friends of the Pact are not required to fulfil specific criteria, and no formal application is needed. They are encouraged to create and disseminate ECP-branded communication material (European Commission, 2025b).

Motivations of Climate Pact Ambassadors

Fundamentally, seeking to act as a CPA corresponds to volunteering in community service, a phenomenon extensively studied within an established research field that integrates concepts and theories from business studies, economics, law, political science, psychology, and sociology (Cnaan and Amrofell, 1994; Cnaan et al, 1996; Wilson, 2000; Williams, 2008). This broader literature includes more targeted scholarship focusing specifically on environmental volunteering, with both streams sharing a common interest in understanding individuals' motivations for volunteering.

CPAs volunteer to participate in a particular institutional arrangement for climate governance: the ECP. Both volunteering and participation research recognize the importance of institutional structures in shaping volunteers' and participants' motivations. In this context, it is essential to differentiate between 'invited spaces' and 'claimed spaces'. Invited spaces are created and organized by public authorities, whereas claimed spaces are initiated by non-institutional actors and are self-organized, ranging from social mobilization to physically created spaces (Cornwall, 2000; Gaventa, 2006; Chilvers et al, 2018; Klaever and Verlinghieri, 2025).

The ambassador programme's design within the EU's institutional framework can be regarded as an invited form of participation, where citizens are specifically encouraged to contribute to a predetermined policy agenda through structured and deliberately non-confrontational channels that prioritize consensus-building over contestation. This invitation mechanism serves as a sophisticated method of controlling and channelling civic engagement, based on the fundamental assumption that such participatory processes are not neutral but embody specific normative orientations and institutional preferences (Wynne, 2007).

This perspective aligns closely with the operational reality of the CPA programme, where participation is curated rather than openly accessible. Potential CPAs must submit formal applications for their mandate, strategically providing the European Commission with opportunities to systematically assess how closely applicants' positions, values, and proposed activities align with the overarching aims and defined scope of the EGD.

While the programme fosters meaningful civic engagement and provides valuable opportunities for citizen participation in climate governance at the European level, it operates as a fundamentally top-down participation model that functions within carefully defined, consensus-oriented boundaries. Rather than challenging existing power structures, questioning dominant policy frameworks, or facilitating more transformative forms of environmental activism, it maintains institutional stability (see, for example, Della Porta and Felicetti, 2022).

Another distinctive feature of the CPA programme is its reflection of the EU's multilevel political system's complexity (Hooghe and Marks, 2001; Knill and Liefferink, 2007; Sommermann, 2015). Potential CPAs must apply for their mandate with the ECP Secretariat, positioned within the European Commission. In terms of content, the mandate entails promoting EU-adopted policies aimed at delivering the EGD. However, most actions CPAs can undertake address the local level, because CPAs are primarily expected to promote the EGD within their communities (Tosun et al, 2023b). This disconnect between applying for a mandate with a supranational organization and implementing the ambassadorship at the local or personal level represents another compelling reason for studying what motivates individuals to volunteer for this role.

Finally, as the programme's name signals by referring to 'ambassadors', this type of volunteering involves a highly visible role that differs significantly from private volunteering. All appointed CPAs are prominently featured on a public website, creating public visibility for their work and facilitating extensive network-building opportunities among ambassadors (Tosun, 2022c; Tosun et al, 2023a; Tosun et al, 2023b). This transparency serves multiple strategic purposes: providing meaningful recognition for ambassadors' contributions to climate action, creating institutional accountability for their commitments and deliverables, and offering inspiration and practical guidance for other potential climate leaders considering similar engagement.

However, CPAs' public visibility can also expose them to unwanted attention from supporting and opposing groups, creating potential personal and professional risks. This exposure can generate conflicts with ambassadors' employers, particularly if they work in industries with different climate or environmental stances or competing business interests. The role's public nature may also deter qualified potential ambassadors who prefer contributing behind the scenes or lack resources to manage public scrutiny, while potentially favouring those from privileged backgrounds who feel comfortable with public exposure and possess the personal resources to navigate potential controversies.

Consequently, the public website generates powerful incentives and significant disincentives for volunteering as CPAs, creating a complex dynamic that represents one compelling reason for examining the multifaceted factors that motivate individuals to apply for the programme despite these potential costs. Given these distinctive features of the CPA programme, it is valuable to study the motivational drivers of those who volunteer for it, which constitutes the first research question guiding this study.

Strategies of Climate Pact Ambassadors

The CPA programme fundamentally focuses on individuals and their capacity for climate action, and therefore, approaching it analytically from the perspective of volunteering, which is inherently individual-focused, appears analytically reasonable. Nonetheless, we can build on the valuable insights offered by the broader and well-established literature on environmentalism and how it strategically engages with the complex institutional framework of the EU.

Environmentalism, in this context, refers to the collective action taken by social movements, civil society organizations, and non-governmental organizations (NGOs) with a view to increasing the level of ambition and the effective delivery of climate and environmental policy across multiple governance levels. Environmentalism in the EU manifests through two main and distinct modes of action: activism and advocacy. According to Parks et al (2023), activism comprises a diverse range of 'outsider' strategies designed to influence policy-making processes, including public protests, demonstrations, and other forms of contentious political action. Advocacy is understood as encompassing a range of 'insider' strategies, which means that advocates strategically draw on their established role, privileged access, and close professional contacts within policy-making circles.

The formal institutional mandate for CPAs is derived from the ECP and focuses primarily on involving citizens in discussions in an attempt to create co-ownership, unlock technological and social innovation, and optimize decision making (European Commission, 2020). More precisely, the ambassadorship comprises the promotion of climate- and environmentally conscious behaviours and community initiatives, facilitating knowledge sharing among participants, and fostering professional networking opportunities. However, the ECP's official website explicitly mentions that one of the tangible benefits of becoming a CPA is to 'gain visibility for your climate action and advocacy on the Pact website and the Pact's social media channels' (European Commission, 2025b). Consequently, while the ECP ostensibly focuses on the concrete delivery of the EGD on the ground by improving decision making and practical implementation, it equally acknowledges and recognizes that CPAs could potentially leverage their official mandate for the broader purpose of policy advocacy (Tosun, 2022c). From this institutional design, it follows that it appears particularly instructive to systematically assess to what extent CPAs actively seek to engage in climate and environmental policy advocacy beyond their formal implementation role.

Being a form of invited participation (Cornwall, 2000; Gaventa, 2006), the ECP is not structurally designed as a venue for contentious forms of individual political action, which corresponds to one prominent form of environmental activism as identified by Parks et al (2023). However, CPAs who do not have direct access to policy makers and, therefore, cannot effectively use insider strategies for influence, can still strategically employ non-contentious, collaborative forms of outsider strategies to indirectly influence policy-making processes and outcomes.

Consequently, the second research question focuses specifically on whether CPAs actively seek to influence climate and environmental policy beyond their implementation mandate, and which particular strategies they employ for advocacy, including non-contentious forms of activism, within the constraints of their institutional framework.

Experiences of Climate Pact Ambassadors

Given that the ECP represents a relatively new and innovative engagement tool, and one which the European Commission has modified and refined over time in order to enhance its effectiveness and broaden its impact (see Chapter 2 and; Tosun et al, 2023a; Tosun et al, 2023b), with the third research question the comprehensive quality of experience that CPAs have accumulated during their tenure is systematically assesssed. This assessment of whether the CPAs considered their experience with the mandate as a positive or negative one is conducted not only to understand the current state of the programme but also with a specific view to identifying concrete suggestions and actionable recommendations for further improving and optimizing the scheme's design, implementation, and outcomes.

Learning about and thoroughly understanding the multifaceted experience of CPAs is of great importance for several interconnected reasons, not least of which is ensuring that the scheme will continue to successfully attract a substantial number of qualified, motivated, and competent candidates who can effectively fulfil the programme's ambitious objectives. This aspect becomes particularly critical when considering the fundamental nature of the ambassador programme as volunteered work, where participation is based on intrinsic motivation and civic commitment rather than external financial incentives (Wilson, 2000; Tosun, 2022c; Tosun et al, 2023b; Marqués Ruizen, 2024).

Indeed, the voluntary nature of the programme means that the European Commission does not offer any direct financial compensations, monetary rewards, or material incentives for CPAs (Marqués Ruizen, 2024; European Commission, 2025b), making the quality of the experience itself, along with the perceived value and impact of their contributions, important factors in maintaining participant satisfaction and the programme's long-term sustainability. Understanding how CPAs perceive their role, the challenges they encounter, the support they receive, and the personal and professional benefits they derive from their participation becomes essential for ensuring continued engagement and for attracting future cohorts of ambassadors who can advance the programme's mission effectively.

Structure of this book

This book is guided by three interconnected research questions discussed by the previous sections, which collectively aim to provide a comprehensive understanding of the CPA programme and its participants' experiences. First, what motivates individuals to become CPAs? Second, do CPAs engage in policy advocacy? If CPAs do engage in advocacy, which strategies do they apply? The third question is what experience CPAs have made with their mandate.

These research questions are particularly significant given the unique position that CPAs occupy within the EU's participatory governance landscape, simultaneously representing official EU climate and environmental policy and supporting their implementation while operating within local communities and potentially pursuing independent policy advocacy goals. Understanding these dynamics is necessary to evaluate the effectiveness of the CPA programme and its broader implications for citizen engagement in the EU's climate and environmental governance (Parks et al, 2023; Machin and Tan, 2024). The research questions are formulated to reflect both the uniqueness of CPAs and to connect this analysis with the broader literature on political participation, which asks similar questions about why people join political parties, for instance (see Bale et al, 2019).

To address these research questions comprehensively and systematically, this book is structured as follows. Chapter 2 provides a detailed overview of the CPA programme, examining its institutional design, operational procedures, and candidate selection mechanisms while placing it within the broader context of participatory governance tools and citizen engagement initiatives employed by the EU. This contextual analysis intends to help readers understand how the CPA programme fits within the EU's approach to democratic participation and climate and environmental governance.

Chapter 3 develops the conceptual framework that underpins the empirical analysis, drawing on relevant theories from different disciplines to establish the theoretical foundations for understanding volunteer motivations, advocacy behaviours, and participatory experiences. This framework integrates insights from volunteering research, policy advocacy literature, and theories of multilevel governance to create the analytical foundation of the empirical analysis.

The research presented in this book is based on original empirical data. Chapter 4 provides details on the database and methodology used for the analysis, explaining data collection procedures, sample characteristics, and analytical approach while clarifying how the key variables central to the research questions were operationalized. These methodological explanations ensure that readers can fully understand and evaluate the foundations of the empirical findings.

Chapter 5 presents and discusses the findings of the analysis, systematically addressing each research question while highlighting key patterns, relationships, and insights that emerge from the data. This chapter examines quantitative patterns and qualitative themes, providing a nuanced understanding of CPA motivations, advocacy activities, and programme experiences.

Finally, Chapter 6 offers concluding remarks that synthesize the main findings, discuss their broader implications for governance and citizen participation in the EU, and identify promising avenues for future research that could further advance our understanding of these important participatory mechanisms in the EU's climate and environmental policy.

Summary

The ECP's ambassador programme represents a unique governance innovation because it is operated by a supranational organization: the European Commission. This supranational characteristic fundamentally distinguishes it from traditional ambassador programmes that have been implemented at the local level, such as by cities and municipalities (Søgaard Jørgensen and Pedersen, 2015), and creates unprecedented opportunities and complex challenges that merit careful examination (Tosun, 2022c; Tosun et al, 2023b). The programme's positioning within the EU's multilevel polity introduces distinctive dynamics that affect how CPAs navigate between supranational policy objectives and local- and community-level engagement, creating a novel form of participatory governance that warrants systematic investigation.

This book is guided by three overarching research questions that collectively explore the CPA programme from multiple analytical perspectives, providing theoretical insights and practical implications for climate governance in multilevel systems. The first of these questions examines what motivations drive individuals to volunteer as CPAs, recognizing that understanding these motivational drivers is essential for academic comprehension of civic engagement and practical programme optimization.

The second question investigates whether CPAs engage in policy advocacy, and if the CPAs do engage in advocacy, which strategies they apply. This dual-part question recognizes that ambassadors may extend their influence in order to attempt to shape policy decisions, potentially employing diverse advocacy approaches from grassroots mobilization to lobbying, thereby creating additional pathways for citizen influence on climate and environmental policy.

The third question explores how CPAs have experienced their mandate, examining positive outcomes and challenges encountered during their ambassadorial tenure. This experiential dimension provides insights into programme effectiveness and identifies potential areas for institutional improvement, while also revealing how participants navigate the complex role expectations inherent in representing EU climate and environmental policy at the local level and among members of their community and people with whom they have close personal relationships.

Overall, this book seeks to advance our understanding of the diverse forms of environmental and climate action, examining who pursues such activities and why they choose to do so, while also considering how institutional contexts shape these choices and constrain or enable different forms of engagement. While the analysis focuses specifically on CPAs, this book aims to generate insights that extend beyond this particular group to provide a foundation for understanding the broader potential outcomes of citizen involvement in climate and environmental action, including implications for democratic legitimacy, policy effectiveness, and the evolution of participatory governance mechanisms in complex political systems.

Effective responses to climate change and environmental degradation undoubtedly require adequate public policy frameworks that address mitigation and adaptation challenges across multiple scales and sectors. However, public policies must be successfully implemented to achieve their intended effects, which directs attention to citizens and their willingness to modify their behaviour in accordance with policy requirements and support broader societal transformation. This implementation challenge is particularly acute in the climate domain (Boasson and Tatham, 2023; Fransen et al, 2023; Boasson et al, 2025), where individual and collective behavioural change is essential for achieving policy objectives, making programmes like the ECP potentially valuable tools for bridging the gap between policy formulation and implementation while fostering the social acceptance necessary for ambitious climate action.

The ECP's effectiveness in facilitating this bridge between policy and implementation depends partly on how well it mobilizes citizen capacity and channels it towards productive action, making this investigation of ambassador motivations, activities, and experiences explicitly relevant to broader questions of climate governance effectiveness.

By examining the ECP through these multiple analytical lenses, this book contributes to the academic understanding and practical improvement of citizen engagement mechanisms in climate governance, offering insights that may inform the design of similar initiatives at various levels of governance while advancing theoretical knowledge about the intersection of participatory governance, climate and environmental policy, and multilevel political systems.

TWO

Climate Action Through the Lens of the European Climate Pact

Introduction

The ECP ambassador programme represents a novel format within the EU context, although ambassador programmes have long existed in other domains. Understanding this innovation requires examining how the fundamental concept of ambassadorship has evolved and has been adapted across different fields.

The concept of 'ambassador' is most familiar from international relations and diplomacy, where the Vienna Convention on Diplomatic Relations of 1961 defines ambassadors as diplomatic agents authorized to represent their home government's interests to the head of state and government of the receiving country. At its core, the convention establishes ambassadors as individuals given a mandate to speak or act on behalf of another entity through external communication and advocacy, a function that transcends the diplomatic realm and provides the conceptual foundation for contemporary ambassador programmes in specific policy domains.

Building on this diplomatic origin, the ambassador concept has found application in policy research, albeit in limited instances. Porto de Oliveira (2020), for instance, introduced the notion of 'policy ambassadors', individuals engaged in promoting policies at different levels of the political system. Related to this are 'cause ambassadors', a category found across various literatures, including sustainability governance research. Cities and municipalities have similarly appointed climate ambassadors, as seen in Denmark (Søgaard Jørgensen and Pedersen, 2015) or at the national level in the United Kingdom (Nadkarni et al, 2019).[1] At the global level, the United Nations Framework Convention on Climate Change (UNFCCC) launched an ambassadorship programme within the framework of the Race to Zero campaign, which is presented on its website as the world's largest coalition of non-state and substate actors taking immediate action to halve global emissions by 2030. The members of Race to Zero are encouraged to compete against each other in developing and implementing climate action towards meeting the goals of the Paris Agreement. The ambassador programme of Race to Zero is called Global Ambassadors. The ambassadors are individuals who are committed to facilitating transformative change. Their mandate is to amplify the campaigns, anchor their narratives locally, and accelerate all types of climate action.[2]

Cause-based ambassador programmes may also target specific demographic groups, such as youth, or focus on areas like cultural promotion through culture ambassadors. Beyond policy and cause promotion, ambassadors serve commercial purposes as well (Reas et al, 2023; Hassler et al, 2025). Companies appoint brand ambassadors to represent and actively promote their brands, from celebrities and influential figures to ordinary citizens who are appointed as ambassadors to promote, for instance, tourism destinations and enhance place image and attractiveness (Rehmet and Dinnie, 2013).

The overall architecture of the EGD reveals why the ambassadorship concept – termed precisely in this way – forms the strategic core of the ECP. Designed as the internal dimension of the EU's comprehensive governance strategy, the ECP directly engages EU member states in transformative action. This domestic focus is strategically balanced by an external dimension that positions the EU as a leader in international cooperation, championing an ambitious 'green deal diplomacy' on the global stage (European Commission, 2019).

Our analytical perspective on the ECP's ambassador programme corresponds to what Hoff and Gausset (2015) regard as a 'governance technology' based on citizen engagement. Such programmes are initiated by public or private agencies, in this case, the European Commission, which asks ambassadors to promote behaviour change among organizations, community members or the citizenry. They serve as tools that stimulate civic engagement and participation. Understood in this way, the ECP's ambassador programme complements the EU's repertoire of climate and environmental governance (Dupont et al, 2024; Kulovesi et al, 2024).

This chapter first presents the ECP's ambassador programme in detail, examining its structure, objectives, and operational mechanisms. Because the ECP emphasizes strengthening democratic participation for delivering the EGD (Çelik, 2025), existing mechanisms for citizen participation in EU climate and environmental policy are reviewed (Machin and Tan, 2024). Finally, how the ambassadorship programme relates to other participation formats is examined, identifying similarities and differences. This analysis provides the empirical background for the subsequent individual-focused investigation, which explores how citizen ambassadors navigate their roles and contribute to climate governance, thereby advancing our understanding of contemporary participatory mechanisms in EU climate and environmental policy.

Features of the ambassador programme

The ECP offers a formal ambassador programme, which follows certain rules. First, an organized selection process exists where potential ambassadors must meet specific eligibility criteria. Second, the programme establishes clearly defined roles and responsibilities for ambassadors. Third, ambassadors receive institutional backing. Finally, the programme incorporates mechanisms for coordination and accountability (European Commission, 2025b). This formal structure ensures that the ambassador programme operates as a coherent governance instrument rather than a loose collection of individual activists or advocates.

At the political level, the European Commission oversees the ambassador programme and the broader ECP initiative (European Commission, 2020). Within the European Commission, responsibility initially rested with Executive Vice-President Frans Timmermans, who led the EGD portfolio until 2023. When Timmermans stepped down to run as a candidate in the Dutch parliamentary elections, Wopke Hoekstra assumed this responsibility (Otjes and de Jonge, 2024). Hoekstra continues to oversee the ECP and the ambassador programme in the second von der Leyen Commission, now serving as Commissioner for Climate, Net Zero and Clean Growth.

At the operational level, the ECP Secretariat coordinates the programme's day-to-day activities, including disseminating information on EU climate policies to the ambassadors and communicating ambassador activities back to EU institutions and other stakeholders. The programme's implementation is further supported by national coordinators in each EU member state, who serve as the primary point of contact for CPAs within their respective countries. These coordinators provide guidance to ambassadors and facilitate networking opportunities and capacity-building activities among the ambassador community (Marqués Ruizen, 2024).

The European Commission's website on the ECP presents a comprehensive list of eligibility criteria and a description of the CPA mandate (European Commission, 2025b). To become a CPA, interested individuals need to submit their formal application to a public call, which is disseminated through many different channels. The appointment is for an initial 1-year period, with the opportunity to apply for renewal each subsequent year. To renew their mandate, CPAs must indicate in their annual reports that they have complied with the requirements of their role.

To be eligible for the mandate, candidates must demonstrate leadership of organizations or communities (whether formal or informal) or of grassroots groups or movements. Influencers and opinion leaders are invited to become CPAs, as are individuals representing cultural, educational, or research institutions. Public office holders (such as members of parliament, civil servants, and mayors) can also become CPAs. On the dedicated website, the European Commission directly invites individuals who have experience in activities related to social inclusion, poverty eradication, and support for vulnerable groups to apply for the programme.

Only residents of an EU member state are eligible for the mandate. Additional eligibility criteria include the aspirant's commitment to undertake at least three concrete actions throughout the year, to complete an e-learning onboarding course within 3 weeks, to participate in at least two ECP Secretariat and two country coordinator-led Pact Community events during the mandate, to maintain regular contact and respond to requests from the country coordinators and the ECP Secretariat, to meet annual reporting requirements as part of the ambassadorship renewal process, and to follow the ECP Secretariat's guidelines when carrying out their activities.

The candidates are also expected to uphold and embody the core values of the ECP throughout their work and engagement. These values serve as guiding principles to foster meaningful climate action and cooperation across the EU. The first value refers to demonstrating a strong foundation in science, responsibility, and commitment. The second value is about embracing transparency and knowledge sharing. Third, CPAs must show ambition and urgency when tackling climate challenges. Fourth, the ECP emphasizes local action and impact, encouraging candidates to engage directly with their communities. The fourth value is that candidates for the ambassadorship are expected to promote diversity and inclusiveness, ensuring that all voices are heard and respected (European Commission, 2025a).

The European Commission further expects the CPAs to engage with the ECP community online groups at least every quarter, to share ECP's messages and campaigns via their networks and communication channels, to share the ECP's achievements via social media, the ECP Secretariat's communication channels, and/or via CPAs' visibility campaigns, and to take part in local, national, regional, or EU-organized events, with the support of their country coordinator or the ECP Secretariat.

The website also formulates clear boundaries for the mandate of CPAs (European Commission, 2025b). CPAs are not authorized to claim that they work for, represent, or speak on behalf of the European Commission or the ECP Secretariat. CPAs do not receive any financial support for their mandate. Instead, they invest time, energy, and money into their ambassadorship (Marqués Ruizen, 2024). They must also not use their mandate to seek undue personal or commercial advantage or use it for greenwashing purposes (European Commission, 2025b).

However, the programme also comes with benefits as highlighted by the European Commission. These include the official recognition of the ambassadorship and the publication of a public profile on the ECP website. The profiles published on the ECP website provide information on the thematic areas the CPAs work on. Initially, the ECP prioritized actions focused on green areas, green mobility, efficient buildings, and training for green jobs (European Commission, 2020). The thematic areas are reported through standardized keywords, with 'climate education and awareness raising' being the most frequently mentioned thematic area. However, there are also other, more specialized thematic areas.

Figure 2.1 shows an overview of the areas in which the CPAs work, according to the information they provide on the ECP's website. Data from 891 CPAs, taken from Tosun et al (2023a), shows that a great majority of them are committed to providing climate education and raising awareness of climate change and

Figure 2.1: Areas in which the CPAs work

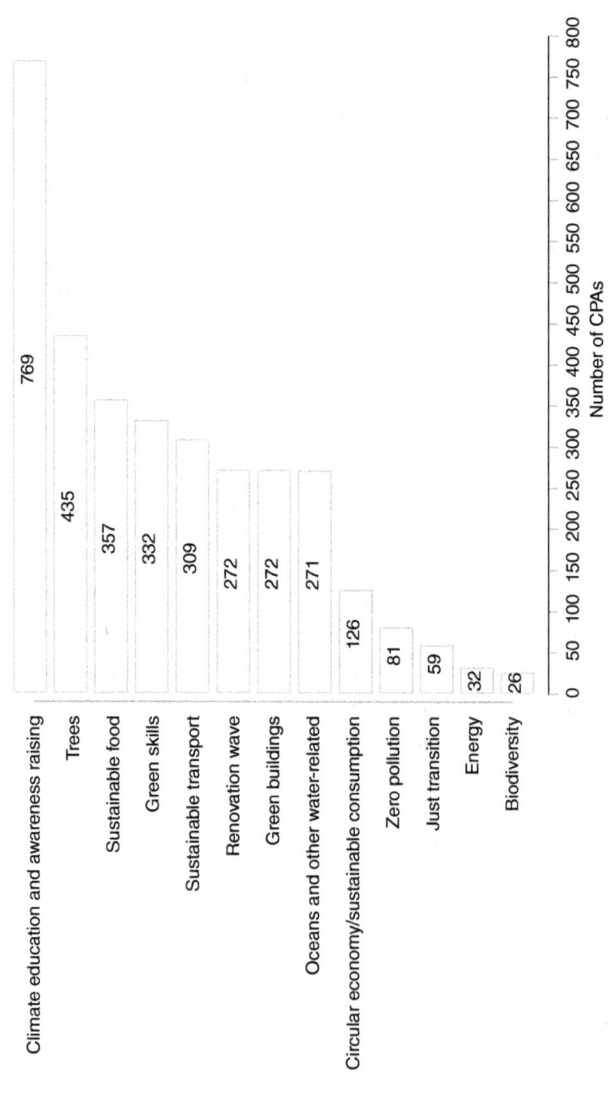

Source: Own elaboration based on data from Tosun et al (2023a)

environmental degradation. The second and third-most stated areas concern activities focusing on trees and sustainable food production. Green skills and sustainable transport follow these areas as the foci of the CPAs' activities.

Given that the ECP aims to facilitate a just transition (Sandmann et al, 2024), it is surprising to see that only 59 CPAs indicated that they regard this as their area of activity. This is worth noting because the European Commission, in its Communication on the ECP, establishes a connection between the ECP and just transition (European Commission, 2020). As with any change to the policy status quo, the delivery of the EGD will create winners and losers. The concept of just transition focuses precisely on this aspect and calls for addressing the social implications of the transformation process—for example, for people working in carbon-intensive sectors—in such a way that does not endanger the achievement of carbon neutrality (Arabadjieva and Bogojević, 2024).

Furthermore, the European Commission stresses that CPAs have access to a pan-European network of CPAs as well as to resources and toolkits on climate action and ECP-branded material for online use. CPAs also benefit from opportunities to gain visibility for their climate action and advocacy on the ECP website and the ECP's social media channels, as well as for participating in ECP-related events across Europe. Furthermore, CPAs have access to online and in-person workshops with policy experts on climate action-related topics and benefit from ad hoc capacity-building support (European Commission, 2025b). This capacity building is delivered by the EU Climate Action Academy, which provides specialized training on selected topics. At the time of writing, the Academy has offered two courses: one focused on engaging citizens in climate action and promoting behavioural change, and the other on developing effective climate communication strategies.[3]

Even though the ambassador programme was only introduced in 2020, in August 2022, the European Commission modified it by formulating more restrictive eligibility criteria, including the requirement for CPAs to be a resident of an EU member state (Tosun et al, 2023b). Previously, the CPAs could reside in any EU and non-EU country, as reflected, for instance, by the data presented by Tosun et al (2023a). The new eligibility criteria deviate markedly from the previous ones since now not even interested individuals from associated countries (for instance, accession candidates, members of the European Economic Area, and Switzerland) can obtain such a mandate.

Another modification concerns the application procedure. Initially, CPA aspirants could apply at any time to become part of the programme through the ECP webpage (European Commission, 2020; Marqués Ruizen, 2024). The CPA aspirants can now only apply during a limited period. In 2024, the call for applications was open from 1 to 31 October. It was widely distributed by EU organizations as well as NGOs and other civil society organizations working on climate and environmental issues.

Furthermore, the European Commission has defined additional requirements and formulated clear expectations for the activities of CPAs. These guidelines also specify the type of action the European Commission wants CPAs to carry out in their roles. The expected activities encompass a broad range of engagement strategies, including giving talks or presentations to diverse audiences, providing education and training on climate-related topics, organizing events such as workshops or community gatherings, undertaking communications activities across various platforms and media, and engaging in policy advocacy to influence decision-making processes at various levels of the EU's multi-level political system. This structured approach ensures that CPAs can effectively contribute to climate action across multiple domains while maintaining alignment with the European Commission's overarching objectives which are defined by the EGD.

The ambassador programme's modifications are noteworthy because they underline its experimental nature. The European Commission and the ECP Secretariat started the programme with an institutional design, which proved to be inappropriate at a certain level. Consequently, they changed the original design, demonstrating that they are willing to learn from their experience. The current programme represents a more developed version, with the EU's ownership reflected in the CPA eligibility criteria, but also regarding the formal definition of the role of the CPAs to which we turn now.

The ambassador programme encompasses a comprehensive range of activities aimed at advancing climate action through diverse channels and outreach strategies targeting different audiences and groups. At its core, the programme is built around the mission of increasing public awareness of the causes, consequences, and urgency of climate change and environmental degradation. CPAs are expected to inform and engage their communities in meaningful dialogue about climate challenges and potential solutions; therefore, reflecting the amply supported insight offered by the scientific literature that pro-environment and pro-climate behaviour cannot be stimulated by information alone (Nadkarni et al, 2019). Therefore, the approach encompasses dialogue and collaborative processes alongside the promotion of sustainable practices and the encouragement of proactive behavioural changes amongst individuals, organizations, and institutions within the CPAs' personal and professional networks.

In doing so, the programme seeks to leverage the ambassadors' existing relationships, social capital, and spheres of influence to maximize its reach and impact. CPAs may organize events, deliver presentations, collaborate with local stakeholders, or produce content for digital platforms, all with the goal of fostering greater understanding and mobilizing collective climate action across sectors and society. By combining communication, advocacy, and community-building efforts, the programme supports the broader objectives of the EGD. It reinforces the idea that everyone has a role to play in the transition towards a more sustainable future.

CPAs serve as active representatives of the ECP by participating in public and private events in their official capacity and communicating about their climate initiatives alongside broader ECP activities. This representative function extends to organizing various engagement opportunities, including online and offline events, discussions organized in formats such as Peer Parliaments (see Chapter 5), and practical climate and environmental activities, from tree planting and organizing and participating in car-free days to information and awareness campaigns.

Beyond direct action and communication, CPAs play a networking and knowledge sharing role. They connect local climate activists with relevant networks and stakeholders, facilitating broader collaboration and resource sharing. The programme emphasizes peer-to-peer learning as a mechanism for knowledge transfer. At the same time, ambassadors also serve as testing grounds for climate solutions that can subsequently be replicated and scaled across different contexts.

This multifaceted approach positions CPAs as implementers of the ECP and EGD-related measures and innovators, combining awareness-raising, direct action, networking, and solution development to maximize their impact across diverse communities and contexts. The activities CPAs are expected to carry out align with the logic of this venue as an invited space (Cornwall, 2000; Gaventa, 2006). While the programme does not ban CPAs from participating in protests and other forms of contentious activism, the list of activities provided on the European Commission's website, as well as the Communication of the European Commission (2020) on the ECP, indicate that the actions shall focus on spreading awareness and supporting action. The ambassadorship is designed as a platform for collaborative action and not for organizing protests.

Citizen engagement in the European Union

The EU and the organizations preceding it have been eager to facilitate citizen involvement through participation and, later, also deliberation. This commitment to democratic engagement has evolved significantly over the decades, reflecting changing expectations about citizen participation in European governance (Bouza Garcia, 2010). In line with the conceptualization of different spaces for citizen engagement, all venues created by the EU correspond to invited spaces (Cornwall, 2000; Gaventa, 2006).

Kohler-Koch (2015) regards the European Economic and Social Committee established in 1958 as the first attempt to facilitate the participation of civil society groups. This pioneering tool aimed at including the perspective of civil society organizations, but even more so that of groups representing the interests of employers and employees as well as other business actors in order to increase the output legitimacy and efficiency of policy making (Quittkat and Finke, 2008).

Alemanno (2021) considers the right to petition as the oldest participatory instrument of the EU. The author dates the origin of this instrument back to 1953 when the Common Assembly of the European Coal Steel Community recognized itself competent and willing to receive petitions. This competence was subsequently assumed by the European Parliament following the creation of the European Communities, with petitions now handled through a dedicated petitions committee.

The landscape of participatory tools expanded further during the 1980s and 1990s. Since the late 1980s, citizens have benefitted from the right to complain to the European Commission regarding breaches of EU law by the member states, including the EU's environmental law (Krämer, 2009). This development marked a shift towards individual citizen engagement beyond organized groups, which was continued and strengthened by the Maastricht Treaty of 1992 and its introduction of European citizenship (Machin and Tan, 2024).

The European Commission introduced the Civil Dialogue in human rights, social justice, the environment, and democracy (Quittkat and Finke, 2008). It is a process in which the EU engages with civil society organizations such as NGOs (Bouza Garcia, 2010). In 1995, the institution of the European

Ombudsman was created as a venue for EU citizens to complain about EU institutions, bodies, offices, and agencies and, therefore, to practise their rights of transparency and accountability (Kostadinova, 2015).

The next generation of participatory instruments was developed in the context of the Commission's White Paper on European Governance (European Commission, 2001). In the White Paper, the European Commission (2001: 7) states that

> the quality, relevance and effectiveness of EU policies depend on ensuring wide participation throughout the policy chain – from conception to implementation. Improved participation is likely to create more confidence in the end result and in the institutions that deliver policies. Participation crucially depends on central governments following an inclusive approach when developing and implementing EU policies.

This vision of participatory governance gradually translated into concrete institutional reforms within the EU. Grounded in a new approach to democratic participation, the European Commission began to implement mechanisms designed to broaden access to policy-making processes. One of the key instruments introduced under this participatory turn was the use of public consultations, which significantly expanded the range of civil society organizations involved in shaping EU policies. This development occurred alongside other participatory mechanisms such as stakeholder conferences and discussions in closed policy (Binderkrantz et al, 2023). The consultations also opened participation to individual citizens, thereby democratizing access to policy discussions and incorporating a more diverse set of perspectives (Quittkat and Finke, 2008).

In practice, the vast majority of these consultations are conducted online, making them more accessible to a broader public across the EU member states. This format reduces logistical barriers and supports large-scale engagement from geographically and demographically varied stakeholders.

A notable example of this approach can be seen in the lead-up to the launch of the ECP. Before officially establishing the initiative, the European Commission organized an online consultation specifically focused on gathering public feedback on the proposed format and aims of the ECP. According to the European Commission (2020), over 80 per cent of respondents expressed strong interest in this participatory model, signalling public support for more inclusive and action-oriented climate and environmental governance.

In parallel with the expansion of participatory opportunities, the European Commission also strengthened institutional transparency through the adoption of Regulation (EC) No 1049/2001, which granted citizens the formal right to request access to documents held by EU institutions. This regulation marked a significant step forward in the EU's institutional openness and accountability. By making the workings of EU institutions more transparent, the regulation aimed to create the necessary conditions for informed and effective public participation (Alemanno, 2021). Transparency, in this context, is not just a procedural ideal but a prerequisite for meaningful democratic engagement: it enables citizens and civil society actors to monitor decision making, scrutinize institutional behaviour, and hold policy makers to account.

It is worth noting that the draft Constitutional Treaty of 2003 incorporated a Principle of Participatory Democracy (Kohler-Koch, 2015). Although this constitutional framework was not implemented due to negative referendum outcomes in France and the Netherlands, its participatory aspirations found expression in subsequent treaty revisions. The Lisbon Treaty, which entered into force in 2009, also includes provisions for participation in Article 11. Paragraph 1 of this article states that the 'institutions shall, by appropriate means, give citizens and representative associations the opportunity to make known and publicly exchange their views in all areas of Union action'. Paragraph 2 states the EU institutions' commitment to interacting with civil society via dialogues, and Paragraph 3 reiterates the importance of broad consultations. Particularly noteworthy is Paragraph 4, which introduces a key innovation in the EU's toolbox of participatory instruments: the European Citizens' Initiative (ECI).

The ECI represents perhaps the most ambitious attempt to create direct citizen influence on EU policy making. It is a process allowing European citizens to call on the European Commission to take action. While ECIs typically request new legislation or amendments to existing laws, some call for the European Commission to refrain from action such as in the case of the 'Stop TTIP' Initiative which called on the European Commission to stop negotiating the Transatlantic Trade and Investment Partnership between the EU and the US and not to conclude the Comprehensive Economic and Trade Agreement between the EU and Canada (Tosun, 2022b; Tosun et al, 2022).[4]

Since ECIs address the European Commission and may result in legislation, they must comply with EU Regulations No. 211/2011 and 2019/788. The operational framework of the ECI reflects democratic aspirations and practical constraints. An ECI begins with an idea for EU legislation. Organizers develop policy demands rather than propose specific legislation. If an ECI succeeds and the European Commission decides to act, these policy demands must be translated into concrete legislative provisions.

The process follows several carefully structured steps (Tosun, 2022b): First, organizers must establish a citizens' committee comprising at least seven EU citizens who reside in seven different member states. Second, the European Commission conducts an eligibility check to ensure the initiative falls within EU competences and aligns with EU values. The European Commission maintains the right to refuse registration. Following successful registration, the initiative enters its mobilization phase. If the initiative passes the eligibility check, organizers have 12 months to collect 1 million signatures from at least seven member states, with each state requiring a minimum number of signatories. Fourth, national authorities verify the collected statements of support. Fifth, the Commission examines the ECI and publishes it in the ECI register. On collecting 1 million signatures, organizers gain the right to meet with European Commission representatives and present their proposal at a European Parliament public hearing.

However, the ECI's democratic potential faces institutional limitations. In the final step, the European Commission decides whether to pursue legislative action. Importantly, the European Commission is not obligated to propose legislation even for successful initiatives. It may instead implement alternative measures to address the ECI's goals, such as recommending organizers submit their initiative to the European Parliament's Petitions Committee.

Climate change has emerged as a prominent theme in citizen-initiated participation. The European Commission provides information on current and past ECIs on a dedicated website.[5] To date, 17 ECIs have demanded policy action from the Commission for tackling climate change directly or indirectly. An exception is the ECI entitled 'Suspension of the EU Climate and Energy Package', which urged the European Commission to suspend the 2009 EU Climate and Energy Package and further climate regulations until a climate agreement was signed by major carbon dioxide emitting countries (Hedling and Meeuwisse, 2015).

A particularly significant example demonstrates the potential and limitations of citizen participation in climate governance. The social movement Fridays for Future launched an ECI calling on the European Commission to acknowledge the climate emergency and take action to limit global warming to 1.5°C. This initiative presented four main demands that illustrate the scope of citizen-driven climate policy proposals.

First, the ECI called on the European Commission to revise its Nationally Determined Contribution under the Paris

Agreement (adopted under the UNFCCC), seeking to reduce GHG emissions by 80 per cent by 2030 and achieve net-zero emissions by 2035. Second, it urged the European Commission to implement a Border Carbon Adjustment mechanism. Third, the organizers demanded that the EU should refuse to sign free trade agreements with countries not following a 1.5°C-compatible pathway according to the Climate Action Tracker.[6] Finally, the ECI requested that the EU provides complementary educational materials for inclusion in member states' school curricula.

Despite the public backing of the Fridays for Future movement, the ECI failed to meet the requirements for success. The European Commission declared the signature collection unsuccessful on 18 January 2022 and consequently did not address any of the organizers' demands. This outcome illustrates the challenges facing even well-organized citizen initiatives in achieving their policy objectives.

Recent innovations have introduced deliberative elements to complement existing participatory mechanisms, marking a significant shift towards more sophisticated forms of citizen engagement in EU governance. From spring 2021 to 2022, the Conference on the Future of Europe took place, which gave citizens an unprecedented opportunity to communicate their views on Europe's future objectives and directions through a comprehensive multilevel process, involving digital platforms, national events, and transnational deliberation (Crum, 2024).

While the conference as such is already a noteworthy innovation with regard to the EU's repertoire of participatory instruments, representing the first major attempt at EU-wide deliberative democracy, it also introduced the European Citizens' Panels as dedicated deliberative forums. Drawing on experiences from EU member states that had already implemented such innovative governance techniques (Lorenzoni et al, 2025), these panels employ a selection mechanism for participants to ensure demographic representativeness. For these panels, around 150 citizens are randomly selected from across all EU member states, carefully balanced to reflect the EU's diversity in terms of geography, age, gender, educational background, and socioeconomic status (European Commission, 2025e).

The deliberative process follows a structured format where panel members meet over three intensive weekends and engage in facilitated discussions about the causes and potential solutions to complex challenges faced by the EU. This format allows for in-depth exploration of issues, informed debate, and consensus-building among participants from different backgrounds and perspectives. Currently, five panels have met and worked out detailed conclusions with specific policy recommendations. One particularly relevant example is the European Citizens' Panel on Energy Efficiency, which discussed current patterns of energy use in the EU and explored how the energy system may need to change in the future to meet climate and sustainability goals. This panel's work demonstrates how deliberative processes can generate nuanced citizen input on highly technical policy areas, bridging the gap between expert knowledge and societal participation in decision making.

Digital infrastructure continues to evolve to support these participatory mechanisms. The European Commission created a single access point for providing information on ways of participation in consultations and other formats, which was previously called Your Voice in Europe. This digital platform was replaced by Futurium in 2021, which was originally developed for discussing digital topics (European Commission, 2025f). In the meantime, the platform is used for engaging with a wide range of topics.

Relationship between the European Climate Pact and citizen engagement

How does the ECP's ambassador programme compare with other existing formats for citizen engagement in the EU? There are indeed notable similarities and meaningful differences between these various participatory formats that merit careful and nuanced examination.

The ECP ambassador programme is fundamentally rooted in the active participation of individual citizens, a trait it shares with several other forms of EU-level citizen engagement discussed in the previous section. However, in contrast to many of these established mechanisms, the ambassador programme places a stronger emphasis on personal initiative, long-term dedication (for instance, renewal of the mandate), and individual responsibility. For example, in the context of the ECI, any EU citizen can support a proposal simply by signing it, with minimal time or effort required. In contrast, the ECP ambassadors commit as individuals but are still expected to engage actively, continuously, and strategically over time, particularly within their communities and networks.

A shared feature across all these participatory mechanisms is that their legitimacy stems from the meaningful involvement of citizens directly or indirectly impacted by EU policy making and governance decisions. Whether through citizen engagement initiatives more broadly or the ambassador programme specifically, the involvement of individuals is typically rooted in voluntary commitment, personal motivation, and alignment with a particular cause or policy area. The ECP ambassador programme, much like other participatory formats, is designed to foster substantive dialogue and create ongoing feedback loops between institutions and communities concerning policy implementation and related challenges about climate and environmental issues. These formats all aim to connect bottom-up engagement with top-down policy processes, ensuring that citizen voices play a meaningful role in shaping outcomes.

Furthermore, all these formats, including the climate ambassadorship, strategically leverage social capital. Personal networks, community-based organizations, professional associations, neighbourhood groups, and even transnational social movements all serve as amplifiers for the reach and impact of citizen participation. This shared reliance on existing relational infrastructure helps embed participatory efforts within communities and enables a broader diffusion of policy awareness and engagement.

Nevertheless, the ambassador programme also presents marked and significant differences when compared with traditional citizen engagement mechanisms. One of the most evident distinctions is the variation in operational scope. While conventional citizen engagement tools such as public consultations, the ECI, or citizens' panels are applicable across the full spectrum of EU policy areas, from digital rights to health policy, the ambassador programme is narrowly focused on the EGD and broader environmental sustainability goals. In that sense, the ambassador role is more policy-specific and mission-driven compared with other forms of citizen engagement. However, general citizen engagement aims to strengthen democratic legitimacy, encourage shared governance, and promote civic dialogue across all domains. The ambassador programme is more targeted in its thematic orientation and strategic ambition: its goal is to deliver the EGD.

Another distinction lies in the nature of commitment and the personal profile of the participants. The ECP ambassadors usually operate within existing institutional and organizational structures to promote environmental and climate-related initiatives. However, their capacity to influence institutional priorities or question systemic approaches remains somewhat limited. Furthermore, the role presupposes a certain level of pre-existing awareness of climate and environmental issues, personal behavioural engagement, and willingness to act as change agents in their personal or professional environments.

This contrasts with other citizen participation formats, which generally do not require such a specialized background. For example, citizens contributing to a public consultation on climate adaptation strategies do not necessarily need prior experience in sustainability, nor must they commit to long-term engagement. In contrast, ambassadors are expected to demonstrate personal leadership, subject matter awareness, and a proactive stance on behaviour change, both their own and that of others within their sphere of influence.

It is also worth noting that prospective ambassadors are unlikely to volunteer for the programme unless they have a strong interest or passion for climate and environmental issues, or regard participation as otherwise beneficial for themselves or for a cause that is close to their heart (see Chapter 3). Traditional citizen engagement tools, on the one hand, remain deliberately open and inclusive, designed to involve citizens regardless of their knowledge level, advocacy experience, or depth of personal commitment to a particular issue.

On the other hand, the ambassador programme provides a unique and valuable shift in perspective by positioning individuals not merely as recipients or audiences of EU climate policy but as co-implementers and agents of change. Through the ambassadorship, the European Commission actively encourages individual citizens to take direct ownership of climate action. Within a clearly defined mandate, ambassadors are empowered to contribute meaningfully to the collective efforts required to realize the objectives of the EGD. Importantly, they are not confined to the ambassadorship role alone; they can also participate in other EU-level formats for citizen engagement, including more specialized governance mechanisms, participatory forums, and advocacy networks; thereby, expanding their influence and scope of action.

Among the stated benefits of the ambassador programme as defined by the European Commission is its capacity to support advocacy. While the primary focus of CPAs is to engage and mobilize their communities, the programme also offers potential pathways for policy advocacy at multiple levels of the EU governance system. As part of their mandate, CPAs may attend local, national, and EU-level events, enabling them to interact with policy makers and institutional stakeholders. In doing so, they gain access to decision-making arenas where they can advocate for more ambitious climate action and support the implementation of the EGD from within.

In addition, even if CPAs do not advocate directly for more ambitious policy measures, they can act as intermediaries, that is, as go-betweens between those who design and adopt policies and the target audiences of these (Abbott et al, 2017). More precisely, they can be regarded as 'climate intermediaries', which is a type of intermediary that focuses on climate issues, which has recently received growing attention in the literature (Tobin et al, 2023b; Tosun et al, 2023c).

Therefore, CPAs occupy a unique and strategic position that enables them to combine various forms of climate engagement effectively. They are able to foster participatory dialogue, inspire and lead grassroots mobilization, and engage in meaningful advocacy efforts that influence policy and societal attitudes. Considering that policy advocacy represents the most direct form of influence on policy making among these engagement formats, this aspect is focused on in the subsequent chapters. In fact, as we have seen, only a few engagement opportunities exist in the EU that facilitate policy advocacy by citizens. Of the engagement formats discussed previously, the one that relates most directly to policy making is the ECI. However, even that instrument does not shape policies directly but rather, in the best of all circumstances, helps to place issues on the political agenda (Tosun, 2022b; Tosun et al, 2022). From this perspective, the ambassadorship offered by the ECP could indeed empower citizens and provide them a means to advocate for more ambitious policies.

Furthermore, given that CPAs are continually supported with up-to-date information, tools, and resources, and benefit from a strong peer network that offers guidance, collaboration, and mutual encouragement, they are exceptionally well-equipped to perform these diverse roles. This ongoing support structure enhances their individual capacity to act and amplifies the overall impact of their initiatives.

As such, it can be reasonably assumed that CPAs possess the knowledge, experience, and commitment needed to act as effective change agents across multiple levels, from local communities to national and European arenas, even if the European Commission (2020) mainly stresses the importance of the local and community level for the CPAs' mandates.

Summary

Ambassador programmes have emerged as a relatively recent and innovative format to involve individuals in climate governance, offering them an additional avenue for engagement and participation. These initiatives are designed to empower people to contribute meaningfully to climate and environmental policy efforts while connecting grassroots action to broader institutional frameworks. As part of the EGD, the European Commission introduced the ECP and the ambassador initiative to help facilitate a just and inclusive transition towards a low-carbon economy. Compared with other formats for citizen engagement, the ECP's ambassador programme represents a very targeted approach in terms of its exclusive focus on climate and environmental action.

However, the European Commission was not the first to adopt such an approach; similar ambassador-style programmes have been developed and implemented by various actors, including international organizations, municipal governments, private companies, and civil society networks (Søgaard Jørgensen and Pedersen, 2015).

At the heart of ambassador programmes lies a central concept: the recruitment of committed individuals who are prepared to engage with the public through a variety of channels, such as social media, public events, and community outreach, in order to raise awareness about climate change, disseminate information about climate-related policies, and inspire tangible climate action. Ambassadors can fulfil several functions, including advocacy for more ambitious climate and environmental action, which represents the focus of this analysis (see Chapters 1, 3, and 5).

The ECP's ambassador programme presents a compelling opportunity for volunteers to actively participate in climate and environment-related initiatives. CPAs are ideally drawn from a wide range of professional fields and social contexts (Marqués Ruizen, 2024). This diversity helps ensure that they can connect with a broad cross-section of the European population; therefore, maximizing outreach and relevance. CPAs are also expected to serve as a feedback conduit for the European Commission and other decision making bodies, sharing insights about how different communities perceive the climate crisis and the urgency for change, which also assigns them the role of climate intermediaries (Tobin et al, 2023b; Tosun et al, 2023c).

It is also worth noting that the European Commission has adjusted the programme's structure over time, even though it has only been in operation for about 5 years. Initially, the programme welcomed aspiring CPAs from EU member states and non-European countries with relatively minimal entry barriers (European Commission, 2019; Tosun, 2022c; Tosun et al, 2023a; Tosun et al, 2023b). Furthermore, the European Commission's expectations of the ambassadors were initially less formalized and specific. Subsequent revisions to the programme's scope and design suggest that the European Commission now expects greater commitment to achieving measurable outcomes, increasing accountability, and aligning the initiative more closely with broader EU climate and environmental goals (Marqués Ruizen, 2024).

While the ECP undeniably enhances opportunities for civic engagement and participatory governance within the domains of climate and environmental policy, it is essential to recognize the limitations of an approach that emphasizes individual action. While not focusing on the ECP but on local-level ambassadorship programmes, critics have pointed out that a heavy focus on personal responsibility can raise ethical and legal concerns, and may even serve as a convenient strategy for political actors to deflect accountability for structural and systemic shortcomings (Hoff and Gausset, 2015).

Therefore, individual action should not be seen as a panacea but rather as one component of a much more intricate and multi-layered approach to reducing GHG emissions, which most fundamentally lies at the heart of any type of climate action (Boasson et al, 2025). From this perspective, the ECP can be understood as a valuable governance tool, one that complements, rather than replaces, top-down policy making. It should also be seen as working in tandem with other grassroots formats, including inter alia, citizens' initiatives at the local and EU-level, all of which contribute to a more democratic and effective response to climate change and environmental degradation.

Regardless of the specific role that CPAs ultimately play, it is important to acknowledge that the ambassadorship is a demanding commitment, particularly in light of the recent changes introduced by the European Commission (Tosun et al, 2023b). The programme now requires more intensive engagement, including enhanced training requirements, increased activity expectations, and more rigorous reporting obligations than in its initial iterations. These heightened demands place considerable time and energy burdens on volunteers, who must balance their ambassadorial duties with their existing personal and professional commitments. The increased expectations underscore the importance of examining what motivates individuals to volunteer for such a role and take on the associated responsibilities.

THREE

Conceptual Framework

Introduction

While the CPA is a new format for citizen engagement in the EU, existing concepts and the corresponding literature can be relied on for exploring it comprehensively. As already outlined in Chapter 1, three key dimensions are of interest: the motivations of individuals to become CPAs, the strategies they use for advocacy, and their experience with the climate ambassadorship. The selection of these three research dimensions is already the outcome of an extensive engagement with the literature. The first one is the literature on volunteering, and the second is the literature on issue advocacy, complemented by research on climate and environmental activism.

Henriksen and Svedberg (2010) note that the literature on volunteering 'tends to be decoupled from the world of political struggle because it is embedded in a discourse of neutral altruism'. The authors call for a more systematic assessment of the possible connections between and across different spheres of civil society, arguing that such artificial boundaries limit our understanding of contemporary civic engagement. This is the first overarching goal of the conceptual framework to be developed in this chapter: to bridge these theoretical divides and create a more integrated understanding of how CPAs operate across traditional boundaries.

While it may sound straightforward to develop such an integrated perspective on volunteering on the one side and advocacy on the other, it is indeed a formidable challenge. This is less due to identifying bridges between the two perspectives, but rather due to ongoing debates within them. For example, due to the complexity of volunteering and the fact that it is subject to dynamic developments, as most social phenomena are, the question of what constitutes volunteering keeps reappearing in the academic discourse. This does not apply to the basic features of the concept but to certain aspects, such as whether individuals seizing one-time or short-term volunteer opportunities can indeed be regarded as volunteers or not (Cnaan et al, 2022). These definitional challenges have important implications for how CPA activities are categorized and understood.

Likewise, it is not easy to determine when individuals act as activists or advocates because, by participating in activism, they can also pursue the goal of influencing policy decisions through various channels. Parks et al (2023) rely on the use of outsider versus insider strategies to differentiate between these two concepts, which conceptually speaking is compelling. Empirically, however, this dichotomy seems too simplistic and does not account for more complex strategic considerations of actors or the fluid nature of contemporary political engagement (Wagner et al, 2023). Furthermore, regarding the ECP, it is possible that CPAs act as activists and advocates simultaneously, depending on the opportunities they are given to demand ambitious climate and environmental action and the specific contexts in which they operate.

The second goal is to explain variation across CPAs in terms of their motivations to join the programme, the strategies they employ for activism and advocacy, and how they perceive the experience they have had with the programme. This analytical objective recognizes that CPAs are not a homogeneous group but rather comprise individuals with diverse backgrounds, experiences, and approaches to the ambassadorship (Marqués Ruizen, 2024). Understanding this variation allows us to move beyond aggregate assessments to appreciate the nuanced ways in which the CPA programme functions across different contexts and participant profiles. This aligns with the main defining feature of the ECP as being a citizen-focused governance technique.

Because the challenge when developing a conceptual framework initially lies with the respective literature on volunteering and advocacy, in this chapter, we engage with both systematically in order to explain how CPAs can be conceptually perceived against the respective literature. We use this review of the conceptual literature to position this study within the broader academic discourse and to explain which conceptual elements have been derived from what literature, thereby establishing the theoretical foundations for the empirical analysis. Subsequently, we introduce some variables that could explain variations among the CPAs with regard to their motivations for volunteering for the ambassadorship, the strategies they use for demanding ambitious policy action, and how they have experienced the programme overall.

This approach allows a move from theoretical conceptualization to empirical operationalization. Regarding the latter, our conceptual framework recognizes the limitations imposed on assessing empirically its validity resulting from the characteristics of the dataset at hand.

Climate Pact Ambassadors as volunteers

One way to think conceptually of CPAs is of them being volunteers and, therefore, drawing from the extensive literature on volunteering (see, for example, Wilson, 2000). Volunteering refers to activities performed with the goal of producing goods and services for others outside the volunteer's household or family. These activities are expected to be performed voluntarily and not because of any type of requirement or obligation. Another important aspect of volunteering is the voluntary giving of time and talents to deliver services or perform tasks with no direct financial compensation expected.

However, volunteers may nonetheless receive some form of support, stipend, or material recognition for their contribution (Cnaan and Amrofell, 1994; Cnaan et al, 1996; International Labour Organization, 2016).

Volunteering can come in two basic forms: formal or organization-based and informal or direct volunteering (Wilson, 2000; Williams, 2008; Lee and Brudney, 2012; International Labour Organization, 2016). Formal volunteering refers to unpaid work that is conducted through established organizations, institutions, or structured programmes. It typically involves an organizational framework or structure, defined roles, responsibilities, and procedures, some form of registration or official recognition, regular commitment or scheduled activities, and it can, but does not have to, include training, supervision, or coordination by the organization. Informal volunteering, in contrast, refers to unpaid help or assistance provided directly to individuals living in other households or communities without going through an organized structure.

From this perspective, participation in the ECP ambassador programme represents volunteering and, more specifically, formal volunteering because the CPAs hold an official mandate for a fixed period and are part of a larger organized initiative with defined procedures and requirements (see Chapter 2). Furthermore, the ambassadorship could be regarded as indicating ongoing volunteering rather than episodic volunteering because the duration of the mandate is above 6 months and the mandate requires certain skills, which the ECP Secretariat aims to provide by offering training and capacity-building formats (see Chapter 2).

Despite the authoritative definition of volunteering given previously, an important discussion within the literature has concerned the empirical identification and mapping of different types of volunteering (Cnaan and Amrofell, 1994; Cnaan et al, 1996; Wilson, 2000; Cnaan et al, 2022). Another important strand in the volunteering literature has focused on the motivations for volunteering, capturing the different types of volunteering, that is, formal versus informal and episodic versus ongoing (Clary and Snyder, 1998; Clary and Snyder, 1999; Ziemek, 2006; Dolnicar and Randle, 2007; Lee and Brudney, 2012; Dunn et al, 2016; Ackermann, 2019). Within this literature, research has also focused on volunteering for advocacy (for example, Nesbit, 2017) and activism (for example, Henriksen and Svedberg, 2010). Furthermore, studies have investigated which factors determine the retention of volunteering, that is, the continuation of volunteering for longer periods (Omoto and Snyder, 2002; Hyde et al, 2016; Merrilees et al, 2020). These literature strands have informed this conceptual framework, and conversely, this analysis and its distinct empirical focus seek to contribute to them.

The engagement with the literature on volunteering allowed the ECP ambassador programme to be identified as a format based on volunteering, which was an important clarification to us. Alternatively, the ambassadorship could have been perceived from a different conceptual perspective, such as that of 'knowledge brokers', which are defined as people or organizations that create connections between researchers and their various audiences (Meyer, 2010). Considering the relevance of research for climate action, this would have been a fitting perspective, but a limiting one since it would have focused on this aspect of the ambassadorship only, that is, the communication and networking aspect of it.

The volunteering perspective is more useful to understand the ambassador programme since, regardless of how the individual CPAs perform their mandate, they must first choose to volunteer for it. In addition, this is the aspect that is perfectly addressed by the dedicated literature on volunteering. Likewise, we contribute to this literature by conceptualizing CPAs as volunteers and shedding light on their personal features and activities.

The choice of focusing on the motivations of individuals to become CPAs also results directly from the discussions in the pertinent literature. Conceptualizations of CPAs from governance studies or policy analysis would not have pointed towards the importance of motivation but rather towards the modes of participation. And yet, the motivation for why individuals participate in engagement formats such as the ECP is important and warrants deeper investigation. Through the focus on the individuals' motivations to volunteer for the ambassadorship, we contribute to the study of climate and environmental governance and policy research in general, and more specifically to the one on climate and environmental policy and governance in the EU. Furthermore, we contribute to the volunteering research literature by addressing the problematic distinction between volunteering and activism/advocacy, recognizing that these forms of civic engagement share overlaps and similarities (Henriksen and Svedberg, 2010).

From this extensive literature, we take the operationalization of the motivations of CPAs as volunteers. Given the size of the corresponding literature, several ways for conceptualizing motivations for volunteering exist, each offering different theoretical perspectives and empirical insights. Considering that CPAs specifically target climate change and environmental degradation, we chose a conceptualization that recognizes volunteering for the production of a public good, that is, a good that is non-rivalrous and non-excludable. More precisely, we used the approach by Ziemek (2006), which postulates three types of motivations for volunteering. The first motivation identified by the author is that volunteers wish to increase the total supply of the public good or service. According to this perspective, the expected benefit of volunteers is an altruistic one, rooted in the desire to contribute to societal welfare and environmental protection.

At the same time, and following Ziemek (2006), our conceptualization acknowledges that volunteering can be driven by the joy of acting as a volunteer and the personal satisfaction derived from meaningful engagement. Based on the private consumption model, volunteering would be driven by the self-value benefit of this activity, wherein individuals derive personal fulfilment and a sense of purpose from their volunteer work. This perspective recognizes that volunteers may be motivated by intrinsic rewards, such as personal growth, social connection, and the satisfaction of making a meaningful contribution to causes they care about.

Third, drawing from the investment model, volunteers may be motivated by their wish to gain labour market experience, skills and also contacts, and seek an exchange benefit from their activity. This perspective acknowledges that volunteering can serve as a strategic investment in one's professional development and career advancement. For CPAs, this might include developing expertise in climate policy, building networks within environmental governance circles, or gaining experience in public engagement and communication that could prove valuable in future career endeavours.

This tripartite framework by Ziemek (2006) thus captures the multifaceted nature of volunteer motivation, recognizing that individuals may be driven by altruistic concerns, personal satisfaction, and strategic career considerations simultaneously. The model's particular relevance to this book lies in its ability to account for the complex motivational landscape that engagement with the ECP represents, where environmental concern, personal values, and professional development intersect.

Volunteering research's interest in the retention of volunteering is again an aspect that governance and policy studies would have either overlooked completely or paid limited attention to. Since governance research tends to focus on the structural level, that is, the governance arrangements and processes, it acknowledges the importance of participation at the aggregate level (for example, Newig and Koontz, 2014; Challies et al, 2017), but it is less attentive to individual-level participation. However, for a programme such as the ECP, the retention of CPAs can be considered relevant for several reasons. One of them is that the EU Commission invests in the training of these people. Thus, if a significant share of

CPAs discontinued their mandate after 1 year, there would be a need for more investments in rebuilding the stock of CPAs and ensuring that they are prepared for their ambassadorship. Another political reason is that a certain level of CPA retention can be interpreted as one dimension for proving the success of the programme. High turnover rates could potentially signal dissatisfaction or inadequate support structures within the programme. From this perspective, volunteer retention has direct implications for both the effectiveness and legitimacy of governance processes and the political success of governance-based measures.

With regard to this research dimension, we found the volunteer process model by Omoto and Snyder (2002) particularly compelling and applicable, as in the second stage of their three-stage model, the authors acknowledge the importance of experience for sustaining volunteering. The authors concentrate on two key aspects of experience with volunteering: satisfaction with volunteer activity and organizational integration. Similar to the previous cases, we contribute to this literature by providing insights into one specific type of volunteering, which, however, can complement the collective insights offered by existing studies while addressing a gap in the understanding of climate policy engagement.

The advocacy perspective

The volunteering perspective offers a compelling lens through which CPAs can be analyzed. However, as argued by Henriksen and Svedberg (2010), the concept of volunteering is not traditionally associated with change-oriented (collective) action. The climate ambassadorship programme, however, aims at achieving precisely this transformative objective: it is one means of meeting the overarching goal of the EDG, which is to bring about transformational changes, not only for climate governance, but also for other environmental sustainability goals, including reducing other forms of environmental pollution and improving biodiversity (Dupont et al, 2024). Therefore, we contend that these conceptual considerations need to be complemented with notions that extend beyond conventional volunteering in order to capture the change orientation that is fundamental to the CPAs' mandate.

Different literature streams across various disciplines offer several concepts. Policy research, for instance, operates with the concept of 'policy entrepreneurs', which can be traced back to the multiple streams framework proposed by Kingdon (1984). In Kingdon's conceptualization, policy entrepreneurs are actors who are willing to invest their resources in the hope of a future return in terms of policy change and societal impact. Petridou and Mintrom (2021) refer to them as 'agents of change' who actively seek policy windows and opportunities for reform. While the concept of policy entrepreneurs resonates with a key feature of CPAs – namely, that they invest their resources to promote climate and environmental action within their communities – Henriksen and Svedberg (2010) refer to a different concept for strengthening the change orientation of volunteering: activism. However, activism is closely related to another concept, which merits consideration, that is, advocacy. Both concepts have been influential in the study of climate and environmental governance (Parks et al, 2023) and both resonate with the broader term of climate and environmental action, which is a multidimensional concept (Tosun, 2022a; Almeida et al, 2024).

Activism involves taking direct and often more confrontational action to trigger change, frequently operating outside traditional institutional channels, with the aim of raising awareness, mobilizing communities, shifting cultural norms, or pressuring entities to change through collective power and grassroots mobilization. Activism can include protests and demonstrations, boycotts or strikes, civil disobedience, and grassroots activities that challenge existing power structures and seek to disrupt the status quo (Tarrow, 1988; Borbáth and Hutter, 2024; Foxe et al, 2024).

Advocacy primarily involves promoting a cause by influencing decision makers, such as politicians, organizations, or institutions, utilizing more formal channels of communication and engagement. It tends to be more strategic, policy-focused, and often works within existing systems and institutional frameworks. It involves action ranging from directly lobbying policy makers and engaging in stakeholder consultations to preparing written policy briefs and organizing information or educational campaigns that seek to inform and persuade rather than confront (Gen and Wright, 2013; 2018).

However, there are important similarities between these concepts that warrant recognition. In essence, both advocacy and activism seek to draw attention to an issue and to increase the number of people who take an interest in it, thereby expanding the constituency for change. Both advocacy and activism aim to influence public policies with a view to initiating or supporting change. However, they may employ different strategies and operate through different channels to achieve these objectives (McKeever et al, 2023).

As discussed in Chapter 2, the ECP ambassador programme can be mostly conceived to align with advocacy, although the target groups of CPAs are not policy makers only, but rather comprise a broader spectrum, including individuals, communities, organizations, as well as decision makers within organizations. This multistakeholder approach reflects the programme's recognition that the delivery of the EGD requires engagement across all levels of society.

Therefore, our analysis is conceptually anchored in the advocacy literature rather than the literature on activism. However, because of the similarities between the concepts, the advocacy literature also captures activism by differentiating between direct and indirect strategies of advocacy. Following Binderkrantz (2008), direct advocacy strategies target either administrative or parliamentary actors, and indirect ones include efforts directed towards the media, and a mobilization strategy where members or citizens are mobilized to take action. In particular, the latter strategy aligns with the notion of activism because it involves engaging and motivating broader constituencies to participate in change-oriented activities. This conceptual framework allows us to examine the CPAs' activities through a comprehensive lens that encompasses both traditional advocacy approaches and more grassroots, activist-oriented strategies within a unified analytical framework.

Thus, we benefit from the literature focusing on policy advocacy by understanding that different groups pursue different strategies and that some of these strategies overlap with activism, such as the staging of or participation in protests or demonstrations (Binderkrantz, 2008; Hojnacki et al, 2012; Dür and Mateo, 2013; Gen and Wright, 2018; Beyers et al, 2020; Wagner et al, 2023). Consequently, even if we draw on the literature on policy advocacy, activism is included when focusing on the full range of advocacy strategies employed by various actors.

Research has demonstrated that direct and indirect policy advocacy strategies should not be viewed as mutually exclusive categories but rather as complementary approaches that can be employed simultaneously or sequentially. In this context, the study by Wagner et al (2023) is worth noting because the authors showed that advocacy groups can embrace both sets of strategies depending on the characteristics of their collaboration networks and the specific policy contexts in which they operate.

More generally, the advocacy literature has informed our conceptual framework by differentiating between different types of direct and indirect strategies (see Chapter 4), thereby providing an apt analytical tool that captures the multifaceted nature of contemporary climate engagement. This theoretical foundation enables the CPAs' activities to be examined through a nuanced lens that encompasses both traditional advocacy approaches and more grassroots, activist-oriented strategies within a unified analytical framework.

The extensive scholarly literature examining advocacy processes and mechanisms predominantly focuses on the strategic behaviours of collective actors, that is, formal organizations and institutional entities, rather than examining the advocacy approaches adopted by individual actors operating within these broader frameworks. This organizational focus reflects the traditional understanding of advocacy as primarily a collective endeavour, where resources, expertise, and influence are pooled to achieve policy change. However, this emphasis on institutional actors leaves a significant gap in the understanding of how individual advocates navigate and contribute to broader advocacy efforts. This is precisely how this analysis contributes to the literature on policy advocacy: by assessing whether individuals acting as CPAs practise policy advocacy and, if they do, which strategies they employ.

Explaining variation at the individual level

This study pursues a dual objective: it aims to describe the differences between motivations, strategies, and experiences among CPAs, and endeavours to provide explanations for these observed variations at the individual level. Given the relatively small number of CPAs who provided complete responses to the questionnaire (as detailed in Chapter 4), the analytical approach must necessarily focus on a correspondingly limited set of explanatory variables. We systematically identified these variables by examining the various bodies of literature consulted during this analysis. We must acknowledge that we did not conduct a comprehensive, full-scale survey specifically dedicated to volunteering behavior. Consequently, we included only a carefully selected, albeit limited, number of items within the questionnaire.

Some of the research we built on to identify the variables that may explain the variation between CPAs has formulated and tested specific hypotheses; this study formulates broad expectations rather than hypotheses and carries out descriptive analyses. This approach reflects the exploratory nature of this investigation into individual-level advocacy behaviour, where the empirical evidence base remains relatively underdeveloped compared with studies on organizational advocacy strategies.

Motivations

The discussion of the variables incorporated into the analytical framework starts by drawing directly from the seminal study conducted by Ziemek (2006), which has served as a fundamental guide in our conceptualization of the dependent variable under investigation. Building on the theoretical foundations established in that work, we identified age and gender as two potentially significant explanatory variables warranting detailed examination.

In an attempt to explain the underlying motivations that drive individuals to volunteer as CPAs, age was considered to represent an important determining factor. Ziemek's (2006) research presents regression results that reveal the statistically significant influence of age on various volunteer motivations. Specifically, her findings demonstrate that advancing age positively influences altruistic motivation, while simultaneously exerting a negative influence on investment and egoistic motivational factors.

Furthermore, her theoretical framework hypothesizes that male volunteers tend to exhibit substantially higher proportions of private consumption and investment motivation, coupled with correspondingly lower levels of altruistic motivation compared with their female counterparts. The empirical findings from Ziemek's (2006) study demonstrate that men are significantly less likely to volunteer for purely altruistic reasons, while they simultaneously show a greater propensity to engage in volunteering activities for egoistic motivations. Although the coefficient generated by the gender variable failed to achieve conventional levels of statistical significance in the altruistic motivation model, it nonetheless exhibits a positive direction that mirrors the pattern observed in the model specifically designed to predict egoistic volunteering behaviour.

We derived the subsequent variables under consideration from the analysis conducted by Tosun et al (2023a), which examined the internet profiles of CPAs and investigated how these individuals rationalize their engagement within the programme. More specifically, the authors rely, among other potential explanatory factors, on the concept of professional identity, which is fundamentally grounded in the professional backgrounds of participating CPAs. The underlying premise suggests that, depending on the respective professional background of individual ambassadors, the motivation to engage actively as a CPA may vary considerably across different occupational categories. Tosun et al (2023a) establish a detailed typology that differentiates between several distinct professional groups: academics, individuals employed within the private sector, members of civil society organizations (including environmental activists), educators, members of the public sector, students, and an additional 'other' category.

In marked contrast to the previously discussed demographic variables of age and gender, we could not draw on extensive existing literature or well-established empirical insights to formulate theoretically-grounded hypotheses regarding professional background effects. Instead, we formulated a broader expectation that professional background can exert a meaningful impact upon the motivational drivers of CPAs, primarily because professional identity fundamentally affects the nature and extent of resources that individuals have at their disposal for programme participation.

Research on volunteering has shown that resources matter when explaining variation in the motivational drivers of volunteering (see, for example, Principi et al, 2016). Being a researcher and having scientific insights into a topic is more likely to result in altruistic volunteering than being a professional working in a company, where motivation may be driven by egoistic or investment reasons. From this perspective, when perceived as a resource endowment, we expect professional background to matter for the type of motivation driving individuals to apply for the ambassadorship.

Beyond the consideration of professional background, an additional novel factor is introduced by Tosun et al (2023a) that might influence ambassadorial motivation and behaviour: the institutional context of representation. Specifically, we assessed whether CPAs act in their individual capacity or serve as representatives of formal organizations. This distinction represents a potentially relevant determinant of motivational patterns because it fundamentally alters the nature of accountability, resource availability, and strategic objectives underlying CPA engagement.

The theoretical basis for this expectation lies in the organization's delegation of representation within the ambassador programme to a member of the organization. In this arrangement, the CPA serves as the agent representing the organization, while the organization acts as the principal. The literature has extensively discussed the principal–agent model, in which authority resides with the principal, while the agent holds an informational advantage. This asymmetry can lead to deviations from the principal's directives (Miller, 2005). While this scenario is conceptually plausible, our focus here is not on such agency problems. Instead, we assume that, compared with CPAs acting independently, those representing an organization are more likely to experience some degree of guidance or direction from that organization, which may affect their motivation for volunteering.

Ambassadors who participate in their individual capacity may be primarily driven by personal convictions, values, and intrinsic motivations related to climate action. Their engagement is likely characterized by greater autonomy in decision making and motivations that align closely with personal climate or environmental beliefs and altruistic concerns. Conversely, ambassadors who represent organizations operate within a fundamentally different motivational framework. Their participation may be influenced by organizational mandates, strategic objectives, and institutional priorities that extend beyond personal environmental concerns. Such ambassadors often benefit from organizational resources, including dedicated time allocation, financial support, and access to networks and expertise. However, they may face constraints imposed by institutional policies, reporting requirements, and the need to align their CPA activities with broader organizational goals.

Similar to the approach regarding the importance of professional background among CPAs, rather than proposing a hypothesis, we maintain the broader expectation that a meaningful difference exists in the underlying motivations of CPAs who serve as representatives of their respective organizations compared to those who engage in the programme acting purely in their individual capacity.

Strategies

While the literature on the motivation for volunteering readily lends itself to formulating expectations about the factors that may explain variation between CPAs, this is less readily the case for variation in strategies. This stems from the fact that CPAs are individuals acting in their own right or on behalf of organizations. However, a large part of the literature on policy advocacy focuses on organizations, that is, collective actors with formalized structures, established resources, and institutional mandates. This is not to say that the literature does not acknowledge the important role of individuals, it does (Gen and Wright, 2013). However, empirical research on policy advocacy still offers a richer body of insights at the organizational level regarding the strategies chosen.

What emerges as theoretically plausible is to focus on two key dimensions that may influence individual advocacy approaches: the professional background of CPAs and the question of whether they hold their ambassadorial mandate in their individual right or serve as formal representatives of established organizations. These dimensions provide an adequate foundation to develop expectations about strategic variation among individual ambassadors.

The professional background of individual ambassadors is likely to serve as a fundamental determinant of whether CPAs can be appropriately characterized as belonging to insider or outsider advocacy groups, drawing upon the well-established theoretical distinction within the advocacy literature (Binderkrantz, 2008). Generally, individuals employed in the public sector—including civil servants, policy advisers, and government officials at various levels—would reasonably be expected to function predominantly as insider groups. Consequently, they would systematically employ insider strategies characterized by direct access to policy-making processes, formal consultation mechanisms, and established institutional channels of influence.

Conversely, we would expect that members of civil society organizations, as well as educators working within academic institutions and students pursuing their studies, would be more likely to belong to outsider advocacy groups. These individuals typically lack direct institutional access to policy-making processes. They are therefore more likely to employ outsider strategies, including public mobilization, media campaigns, grassroots organizing, and other forms of external pressure designed to influence policy indirectly through public opinion and broader societal engagement.

With regard to individuals employed within the private sector, the situation becomes considerably more complex and theoretically ambiguous, making it substantially more difficult to formulate clear expectations a priori. It remains entirely possible that some private sector ambassadors are employed by highly influential multinational companies or industry leaders that possess significant lobbying capacity and established channels of access to policy makers, which would position them closer to insider advocacy approaches.

However, it could equally be the case that many private sector ambassadors work for smaller, less politically influential companies that lack substantial policy access or lobbying resources. The latter scenario should be considered more theoretically plausible given that the ambassadorship specifically focuses on climate action and environmental advocacy, areas where smaller, more progressive companies with limited political influence are likely to be disproportionately represented among voluntary participants, rather than large corporations that might face potential conflicts between their commercial interests and ambitious climate commitments.

As argued previously, the distinction between insider and outsider groups and the corresponding insider and outsider strategies should not be seen as a strict dichotomy. Consequently, while we use this distinction to explain why professional background may influence the choice of advocacy strategies, we also recognize that the empirical reality will be more nuanced. Should this prove to be the case, it would provide additional support for the calls, as expressed by Wagner et al (2023), for example, to overcome the insider–outsider approach in the literature on policy advocacy.

Regarding CPAs who represent an organization, their organizational affiliation likely influences the advocacy strategies they employ, reinforcing the understanding that they must be viewed as agents of their organization (Miller, 2005). However, it is again challenging to formulate a clear hypothesis due to the competing mechanisms at play.

CPAs representing organizations could represent public bodies, which would then give them inside access to policy makers. Therefore, they can be expected to choose these direct advocacy instruments more frequently. However, it is also plausible to reason that CPAs representing organizations are much more constrained in how they deliver their mandate, particularly when bound by institutional protocols and formal accountability structures. Such constraints could result in them focusing on other activities beyond direct advocacy, such as networking and information sharing, for instance. This duality suggests that organizational affiliation may simultaneously enable certain advocacy strategies while constraining others, creating a complex relationship between institutional backing and strategic choice.

Summing up, while the professional background of CPAs provides a useful potential explanatory variable for understanding strategic variation through the insider–outsider distinction, we anticipate the empirical reality to be far more nuanced than our conceptual considerations suggest. The additional consideration of whether CPAs represent established organizations further complicates the picture, because institutional affiliation appears to create opportunities and constraints that may push advocacy strategies in different directions. This complexity underscores the need for more empirical research at the individual level, which feeds into theory refinement or even informs theory building.

Experiences

Based on the volunteer process model (Omoto and Snyder, 2002), we reason that the perception of the ambassadorship could be affected by the initial drivers of the mandate, that is, the motivation. Thus, following our conceptual considerations based on Ziemek (2006), differences could exist between CPAs who chose to volunteer because of altruistic, investment, or egoistic reasons. However, we are not aware of any theoretical basis for formulating a more refined expectation regarding which type of motivational driver could result in a more positive or negative assessment of the ambassadorship experience.

Furthermore, we contend that the experience of the ambassadorship is likely to be closely connected to concepts of internal and external political efficacy, which represent fundamental dimensions of political engagement and civic participation (Bale et al, 2019; Vecchione and Caprara, 2009; Oser et al, 2022). Internal political efficacy refers to individuals' confidence in their ability to understand and participate effectively in political processes, while external political efficacy concerns beliefs about the responsiveness of political institutions and decision makers to citizen input (Bandura, 1997; Beierlein et al, 2012).

Shore and Tosun (2019), for instance, demonstrate how young people's experiences with public employment services can shape levels of external political efficacy, that is, the feeling that decision makers are responsive to citizens' needs. Complementary to this research perspective, CPAs' political efficacy in combination with the actual experience they have gained during the execution of the mandate could explain whether the overall experience is perceived as positive or negative.

With regard to internal political efficacy, CPAs possessing high levels of this efficacy type may perceive the programme positively if they believe it has afforded them sufficient space and opportunity to carry out the types of actions they originally intended to undertake. This perception stems from their confidence in their capacity to influence outcomes and navigate the programme structure effectively. More generally, individuals with high levels of political efficacy should be significantly more likely to feel positive about their overall programme experience, since they maintain a stronger belief in their ability to influence and shape their experience themselves through their actions and decisions.

In contrast, CPAs with high levels of external political efficacy will perceive the programme positively if they are under the impression that their actions and efforts have triggered some kind of meaningful response from the relevant institutions and decision makers. It is conceivable that such individuals are more likely to be disappointed by the ambassadorship programme if they do not receive a tangible reaction or acknowledgement from the institutional level to their own or the CPAs' collective activities, because this would contradict their expectations about the system's responsiveness to citizen engagement.

Summarizing the main elements of the conceptual framework

Table 3.1 presents the conceptual framework that operationalizes the three key analytical dimensions examined in this study of CPAs. The framework systematically organizes the three outcome variables alongside their corresponding explanatory factors across each dimension.

The motivations dimension examines three types of volunteer motivations—altruism, egoism, and investment—as outcome variables, with age, gender, professional background, and mandate type serving as explanatory variables. This reflects the tripartite motivational framework derived from Ziemek's (2006) theoretical approach.

The strategies dimension focuses on direct and indirect advocacy strategies as outcome variables, with professional background and types of mandate as explanatory factors. This reflects the advocacy literature's distinction between insider and outsider strategic approaches (see Binderkrantz, 2008). The

Table 3.1: Conceptual framework

Dimension	Outcome variables	Explanatory variables
Motivation	Altruism Egoism Investment	Age Gender Professional background Types of mandate
Strategy	Direct advocacy strategy Indirect advocacy strategy (activism)	Professional background Types of mandate
Experience	Positive experience Negative experience	Altruism Egoism Investment Internal political efficacy External political efficacy

potential explanatory factors, professional background and type of mandate, are taken from the study by Tosun et al (2023a).

The experiences dimension presents positive and negative experiences as outcome variables, with a dual set of explanatory variables: the three motivational drivers (altruism, egoism, and investment) and political efficacy measures (internal and external political efficacy). This reflects the theoretical expectation that both initial motivations (Omoto and Snyder, 2002) and self-efficacy beliefs influence ambassadorship experiences. Between these two expectations, the second is less well-established in the literature than the first and can therefore make a more meaningful contribution to theory development.

Summary

This chapter has developed a conceptual framework for understanding CPAs by the integration of volunteering and advocacy literatures, addressing the artificial boundaries that have traditionally separated these domains of civic engagement. By bridging these theoretical divides, we established a foundation for examining the motivations, strategies, and experiences of individuals operating within the ECP ambassador programme.

The conceptualization of CPAs as formal volunteers provides important insights into their role within contemporary climate and environmental governance. Drawing on Ziemek's (2006) tripartite framework, the CPA motivations encompass altruistic concerns for climate change and environmental protection, personal satisfaction derived from meaningful engagement, and strategic investment considerations related to career development and skill acquisition. This multifaceted understanding moves beyond simplistic assumptions about volunteering and acknowledges the complex motivational landscape that characterizes engagement in climate and environmental governance.

Equally important is the theoretical positioning of this study within the advocacy literature, which captures the change-oriented nature of the ambassadorship that traditional volunteering frameworks struggle to accommodate. By anchoring this analysis in advocacy theory while recognizing the overlap with activism through Binderkrantz's (2008) direct and indirect strategic categories, we created a unified analytical framework, which is capable of examining both traditional advocacy approaches and grassroots, activist-oriented strategies. This integration is particularly valuable given that CPAs are expected to operate across multiple stakeholder groups, including individuals, communities, organizations, and decision makers, and to use a wide range of strategies (European Commission, 2020).

Our analytical approach recognizes that CPAs represent a heterogeneous group whose variations in motivations, strategies, and experiences warrant systematic explanation. The variables identified for explaining this variation – age, gender, professional background, organizational representation, and political efficacy – reflect careful consideration of both theoretical reasoning and empirical feasibility. While we acknowledge the exploratory nature of this investigation and its data limitations, the variables underlying our conceptual framework provide a solid foundation for understanding individual-level differences among climate advocates.

We expect the empirical focus on individual rather than organizational advocacy behaviour to make a contribution to the advocacy literature more generally. Most existing empirical research focuses on collective actors with formalized structures and institutional mandates, leaving individual advocates relatively understudied despite their growing importance in contemporary climate and environmental governance. Our framework enables the examination of how individual advocates navigate strategic choices, operate within informal networks, and balance personal motivations with broader climate objectives.

The connection established between ambassadorship experiences and political efficacy concepts provides valuable insights into the programme's potential for developing civic capacity and sustaining long-term climate engagement in line with the volunteer process model by Omoto and Snyder (2002).

Nonetheless, we recognize that our conceptual framework would benefit from further refinement. To achieve this, we consider it a promising strategy to apply this framework to real-world empirical data and then revise it based on the resulting observations.

FOUR

Research Design and Operationalization

Introduction

Understanding the motivations, strategies, and experiences of CPAs requires a methodological approach that can capture both the diversity and complexity of this distinct type of volunteering. Our research can build upon established methodological frameworks from the broader volunteering literature. We employ validated measurement instruments that have been successfully applied to other forms of environmental and civic engagement. This chapter outlines the data collection strategy employed to investigate this research and how the key variables were operationalized.

The ECP ambassador programme represents a novel institutional format for engagement, which presents both opportunities and challenges for empirical research. As discussed in Chapter 3, this novelty necessitates careful adaptation of existing theoretical frameworks to capture the distinctive features of supranational volunteer engagement—a requirement that extends equally to our methodological approach.

While the ECP's online platform provides transparency through publicly accessible ambassador profiles, these sources offer limited insight into the underlying motivations that drive individuals to volunteer, the concrete advocacy activities they pursue, or their personal experiences with the programme's effectiveness. These methodological constraints necessitated a more direct approach to data collection by developing and implementing the European Climate Pact Ambassador Survey, administered in November 2023 to the population of 559 active CPAs (Rzepka and Tosun, 2025). Drawing on established theoretical constructs and survey items from volunteering research and policy advocacy studies, the survey was designed to capture three fundamental dimensions: the motivations driving CPA participation, the extent and nature of their policy advocacy activities, and their overall satisfaction with the programme experience. This survey represents the first empirical assessment of climate ambassadors operating under a supranational mandate. These insights have important implications for understanding how supranational institutions can effectively mobilize citizen engagement in climate action.

This chapter proceeds by examining the methodological considerations that shaped the research design, the decisions taken regarding operationalization, and the analytical approach adopted for interpreting the survey findings. While acknowledging the limitations inherent in this database, the methodological framework underlying it can be regarded as a foundation for understanding why individuals volunteer for the ambassadorship programme, whether and how they use the mandate for policy advocacy, and what their assessment of the programme is.

Public profiles of the Climate Pact Ambassadors

A defining characteristic of the CPA programme is its fundamentally people-centred approach. Rather than simply providing neutral information on climate and environmental issues and leaving individuals to interpret and act upon it independently, the ECP deliberately leverages the diverse characteristics of its ambassadors – including their varied ages, genders, professions, and socio-economic backgrounds – along with their interpersonal interactions to make climate change and environmental degradation more relatable (Nadkarni et al, 2019). Through this human-centred strategy, the programme helps citizens develop a clearer understanding of realistic actions they can take in their daily lives to address these pressing challenges.

Consequently, and as outlined in Chapter 2, a key feature of the ECP ambassador programme is that the ambassadors have public profiles. In fact, the working of the programme critically depends on CPAs visibly acting in this capacity and maintaining a public presence, which can already be derived from the title for their activity: the ambassadorship (Reas et al, 2023).[1] A dedicated website presents the CPAs, provides their contact details, portrays their mission statements as well as their personal thematic areas, and their countries of residence.[2] This digital platform serves as the primary interface between the ambassadors and the broader public, facilitating engagement and knowledge sharing across diverse communities.

Of these elements, the mission statements (denoted as 'responsibilities' on the website) are short texts. They are the only ones that are not standardized and demonstrate considerable variation in what the CPAs chose to share with the public. The texts include different types of information and at different levels of detail, from the personal or professional background of the CPAs, their motivations for deciding to volunteer for the programme, how they are already practicing climate action or seek to contribute to climate action taken by others, and whether they act on their own right or behalf of an organization.

This diversity in presentation reflects the varied backgrounds and approaches that different ambassadors bring to their roles, which can be considered a strength of the programme (Marqués Ruizen, 2024), particularly given that the CPAs are designed to facilitate a just transition to the EU's net-zero objective (European Commission, 2020). Such a transition necessitates the inclusion of perspectives from numerous different groups across society, making this diversity not merely beneficial but essential for the programme's effectiveness.

While this ECP website is informative and has served as a valuable source for research assessing the CPAs' commitment to young and future generations (Biesbroek et al, 2018; Tosun et al, 2023a), there are also severe limits to what can be inferred from it. Relating to the variation in the mission statements with regard to their length, structure, and information context, it is difficult to obtain comparable data from this source for systematic analysis. More specifically for these research interests, the text on the website may provide insight into the motivation of some CPAs for volunteering, but certainly not for all ambassadors comprehensively. Therefore, a few statements published in the official CPA online profiles are examples to illustrate the findings of this analysis.

The content of the website is even less helpful for learning about the concrete actions CPAs carry out in their day-to-day work. While some CPAs list different kinds of actions, such as offering courses on climate change or guiding businesses in adopting more sustainable practices, these seemingly represent intentions or planned activities rather than documented actions they have undertaken and completed. Furthermore, the interest of this research lies particularly in strategic actions designed to influence policy decisions, for which the website does not provide an adequate means of communication. This limitation exists both because some strategies still need to be defined and developed, and because it is reasonable not to communicate all of these strategies publicly to ensure their effectiveness and maintain the strategic advantage (Gen and Wright, 2013, 2018). Additionally, the public nature of the platform may constrain ambassadors from sharing sensitive or politically nuanced approaches that could be vital to their work.

Naturally, the website does not contain information on how CPAs rate the experience they have had with their mandate or their assessment of the programme's overall effectiveness in achieving its stated objectives.

Another way to obtain publicly available data for CPAs and their activities is via their social media profiles. Of course, these are less accessible than the public profiles provided on the ECP website because most of these social media channels require registration and account creation. Within this public but more protected space, more could be learned about the CPAs' motivations and especially their day-to-day activities and engagement patterns. However, the data that could be obtained via the social media profiles is again both difficult to compare systematically and suffers from significant methodological limitations.

Based on our investigations for this study, some CPAs are considerably more active on social media than others. This variation can be attributed to several factors: differences among the CPAs regarding the extent to which they embrace the publicness of their mandate, their technical capability or comfort with using social media platforms, their willingness to share personal and professional information online, and their available time for maintaining an active digital presence. Additionally, generational differences and cultural attitudes towards social media usage may influence their level of engagement. Furthermore, the transformation of Twitter into X has complicated efforts to document the complete range of CPA social media activities, since many ambassadors migrated their online engagement to other platforms following these changes.

Likewise, the limitations listed previously hold true regarding the possibility of obtaining the kind of comprehensive information of interest via public social media profiles. Generally speaking, even for those CPAs who are active on social media, we are most likely to learn about the activities of CPAs and their general advocacy efforts (see Chapter 5), but less about their underlying motivations to seek this mandate and their personal experiences with the programme's challenges and successes. Social media content tends to focus on outward-facing communications rather than reflective analysis of their role.

Consequently, to obtain the empirical information needed to answer the research questions comprehensively, we chose a different methodological approach for data collection that would allow for more direct engagement with the CPAs themselves. We deemed this shift in methodology necessary to capture the nuanced perspectives and experiences of the ambassadors, which could not be adequately accessed through sources discussed above.

European Climate Pact Ambassador Survey

While the ECP website was of limited use to directly collect the data necessary for answering the research questions, it nevertheless provided the essential point of departure for this investigation. As already stated previously, the website contains information about the CPAs, including their names and detailed contact information on how they can be reached through various social media platforms or by other means, such as email correspondence. Building on this foundation, we used the contact information to field an online survey, the European Climate Pact Ambassador Survey, which was administered between 8 and 30 November 2023 (Rzepka and Tosun, 2025).[3]

The online survey was programmed in Google Forms and distributed to the population of 559 CPAs as identified on 8 October 2023. It should be noted that the number of CPAs as listed on the website is time-sensitive because it contains only those CPAs who currently hold an active mandate, and the mandates of these individuals commence and terminate on different dates throughout the year. Therefore, the extraction of data at different points in time would inevitably have resulted in a different population of CPAs being captured. This temporal variability is clearly illustrated by Tosun et al (2023a), who based their analysis on a population of 876 CPAs, which corresponds to the numbers extracted from the website on 2 January 2023. At the time of writing, the number of

CPAs had reached even higher levels, with more than 1,100 of them being listed on the dedicated online platform.[4] This increase illustrates the time-sensitive nature of data extraction, highlighting how findings are inherently bound to the chosen temporal framework of the study.

The notable reduction in the overall number of CPAs between 2022 and 2023, and indeed throughout 2023, can be directly attributed to a significant change in eligibility criteria for CPAs (see Chapter 2). Originally, the programme operated under an open access model whereby anyone could become a CPA regardless of their country of residence. This inclusive approach is reflected in the study by Tosun et al (2023a), who differentiate between CPAs based in an EU member state or a country associated with the EU (i.e., accession candidates, members of the European Economic Area, and Switzerland) and other countries. From the latter category, individuals from India, Tunisia, and Nigeria dominated the population of CPAs at the time of data extraction, representing a significant proportion of the international participation.

All 559 CPAs at that time received a personalized invitation to participate in the online survey via the contact details offered on the ECP website. For some individuals, we could obtain a direct email address from their public profiles, whereas others we had to contact via their social media profiles or other communication channels they had specified on the ECP's website. In addition to the direct outreach, the ECP Secretariat kindly sent out an official invitation to CPAs to participate in the survey, which helped to legitimize the research endeavour and potentially increase response rates.

The invitation letter set out the survey's primary goals of learning about individuals' motivation to volunteer as CPAs and whether they have made effective use of the opportunity offered by their mandate to engage in policy advocacy activities. This document clearly explained that no personal information from the CPAs would be used inappropriately and that any responses given by them would only be processed and analyzed in strict accordance with the data protection declaration. The invitation letter also offered respondents the opportunity to receive a comprehensive summary of the findings on completion of the research. Apart from the research summary, we did not provide any monetary or other material incentives to encourage survey participation.

To indicate that the research builds on a broader, well-established research agenda and to give the CPAs a sense of the previous scholarly work in this area, the invitation letter included references to relevant publications, for example, Tosun (2022c), Tosun et al (2023a), and Tosun et al (2023b). These publications engage with CPAs in different ways and from various analytical perspectives. Tosun (2022c) is a descriptive piece and presents a comprehensive overview of the ECP and the ambassador programme's structure, whereas Tosun et al (2023b) contend that CPAs can act in different functions, among which is acting as policy advocates within their respective communities. Tosun et al (2023a) focus specifically on how the CPAs perceive their social and professional roles and whose interests they claim to represent in their advocacy work.

Although CPAs are based across different EU member states, we have implemented this survey exclusively in English, as CPAs are expected to be fluent in English given the transnational nature of their mandate. The invitation letter was also provided in English only.

In total, 78 CPAs completed the survey successfully, of which 61 indicated that they are CPAs in their own right and 17 stated that they represent an organization or institutional body. Consequently, the overall response rate was approximately 14 per cent, which, while modest, provided sufficient data for meaningful analysis.

Compared with the share of organizational CPAs reported by Tosun et al (2023a), the ones included in this survey are somewhat over-represented. In the data collected by the authors, organizational CPAs account for 10 per cent of the total population, whereas in this survey's data, they account for about 20 per cent. This is not a methodological problem per se, because no causal conclusions will be drawn from the analysis. Nonetheless, it is worth bearing this distinctive feature of the survey data in mind when interpreting the results. Given the goal to offer a first explorative analysis of this population, the over-representation of the organizational CPAs has the analytical advantage that there are more observations for this particular group of CPAs.

Given that CPAs display their contact details publicly on the platform and considering that their mandate explicitly involves communication and interaction with various communities, we initially expected a higher response rate. Despite the relatively low response rate, there was sufficient variation in the responses to carry out meaningful analyses across different variables of interest. However, with one notable exception, the small sample size forced us to primarily draw on descriptive statistics and to regard the findings as preliminary and indicative, which means that future research is required to update and broaden the database and to replicate our analysis in order to test the robustness and generalizability of the findings.

The survey implemented includes different types of questions designed to capture various aspects of the CPAs. The first set gathers comprehensive information on socio-demographic characteristics (Questions 1–5) and includes the CPAs' country of residence and birth, followed by their age (mean = 45, range 22–80), the gender they identify with (40 per cent female, 58 per cent male, and 2 per cent non-binary), and their current occupation and professional background. The response categories for the latter were taken directly from the validated classification system used by Tosun et al (2023a), ensuring consistency across studies.

Operationalization of the variables

The outcome variables of interest for this study examine three fundamental dimensions: (1) the motivation to become a CPA; (2) strategies to conduct policy advocacy activities;

and (3) the CPAs' overall experience with the mandate. These variables form the analytical core of the subsequent chapter and will receive more detailed attention there. Consequently, the presentation of the variables here focuses primarily on the precise wording of the questions asked and provides information on the sources consulted in their formulation.

We present a detailed overview of the specific response options in the following analytical section and in the codebook stored in our institutional data repository.[5]

Motivation

The motivation for becoming a CPA should correspond fundamentally to the motivation for volunteering more generally (see Chapter 3 for details). Volunteering is defined as 'any activity which involves spending time, unpaid, doing something which aims to benefit someone (individuals or groups) other than or in addition to, close relatives, or to benefit the environment' (Williams, 2008).

Given that volunteering is of great value to society and contributes significantly to social cohesion, public authorities have actively supported the collection of volunteering-related information through systematic research. These data are typically collected by dedicated surveys on volunteering, such as the Swiss Volunteering Survey (Stadelmann-Steffen and Gundelach, 2015) or the comprehensive German Survey on Volunteering (Kelle et al, 2021), which provide valuable insights into volunteering patterns and motivations.

To start with the first dimension, for the motivation of becoming and being a CPA, established items were borrowed from Ziemek (2006) and adapted partly to suit this specific research context. Question 6 in the questionnaire is worded as follows: 'Think about your decision to become a Climate Pact Ambassador. How much do you agree or disagree with the following statements?' The respondents who selected multiple answers, among which were the following:

- Altruism: 'I volunteer as a Climate Pact Ambassador because of my social principles/moral obligation'; 'I think there is a great need for somebody to do this kind of work';
- Egoism: 'It makes me feel good'; 'I had an interest in the activity or work';
- Investment: 'I want to make new contacts that might help my career'; 'I want to learn new skills and/or receive training'.

To filter and identify the highest priority motive for their commitment and to increase variation, Question 8 required one single answer: 'What do you consider the one most important factor in your decision to become a Climate Pact Ambassador?' This question is a constructed compilation of items based on a survey fielded by the Australian Bureau of Statistics (2021) and Ziemek (2006), designed to capture the primary driving force behind the CPAs' participation in the programme. Even though the wording slightly varies, the answers still compare with those three categories in the previous question:

- Altruism: 'Society's need for committed people'; 'Social or moral principles';
- Egoism: 'It makes me feel good'; 'Interest in the activity or work';
- Investment: 'Community'; 'Use skills and experience'.

Strategy

The second set of survey questions gauges whether CPAs have applied certain policy advocacy strategies. To obtain a specific understanding of potential advocacy activities carried out after being appointed a CPA, we asked two questions that delve deeper into their post-appointment behaviour. The first of them (Question 13) was the following: 'There are different ways of pursuing climate action. Since the beginning of your mandate as Climate Pact Ambassador, have you done any of the following?' The survey participants could choose from a comprehensive list of activities; for each activity, they needed to indicate 'yes' or 'no'. This question is an adapted version of the one used in the core module on Online Political Participation of the 2016 edition of the European Social Survey (ESS ERIC, 2017), which has been validated across multiple European contexts. This list of items was chosen because it also includes types of advocacy that overlap with activism, such as taking part in a demonstration (see Henriksen and Svedberg, 2010):

- 'Contacted a politician, government or local government official';
- 'Worked in a political party or action group';
- 'Worked in another organization or association';
- 'Worn or displayed a campaign badge/sticker';
- 'Signed a petition';
- 'Taken part in a lawful public demonstration';
- 'Posted or shared anything about politics online'.

We took the next question (Question 14) from the established research by Beyers et al (2020) and adapted it to our specific focus, which reads as follows: 'Since the beginning of your mandate as Climate Pact Ambassador, how often have you actively sought access to the following institutions and agencies in order to influence public policies?' The response options come with a scale of frequency, ensuring consistency in measurement across different advocacy activities. For better visualization results, this variable was recoded as a dummy variable, differentiating between 'no' for the answer 'I did not seek access' and 'yes' for all frequencies from 'at least once' to 'at least once a week'. The response options include the following:

- Ministers (including their assistants/cabinets/political appointees);
- Elected members from the majority or governing parties of parliament;

- Elected members from minority or opposition parties of parliament;
- National civil servants working in the Head of State's/Head of Government's Office;
- National civil servants working in departmental ministries, such as agriculture, environment, transport, and health;
- National civil servants working for the coordination of EU affairs;
- Courts.

Experience

To evaluate the third analytical dimension, the experience of the volunteers, this research enquired about the CPAs' satisfaction with their mandate and their overall assessment of the programme. For this purpose, Question 15 asked how the CPAs describe their overall experience with the mandate, ranging from very positive (one) to very negative (five) on an ordinal scale, which was then recoded in reverse for visualization purposes.

Furthermore, when examining the third analytical dimension – the CPAs' experience with the programme – we included a measurement of political efficacy as developed by Beierlein et al (2012), in addition to the motivational dimensions of altruism, egoism, and investment.

The internal political self-efficacy index combines two items: 'I am good at understanding and assessing important political issues' and 'I have the confidence to take active part in a discussion about political issues.' Most CPAs reported high or very high levels of internal political self-efficacy, with 82 per cent selecting four or higher on a scale from one (very low) to five (very high). This observation comes as no surprise: if people were considered to have a low internal political self-efficacy, they would not have volunteered in the first place.

The external political efficacy index is based on two items: 'Politicians strive to keep in close touch with the people' and 'Politicians care about what ordinary people think.' The response options to these four questions utilize a five-point Likert scale, from one (strongly disagree) to five (strongly agree). To calculate the score for internal and external political efficacy, the mean of the responses to the two questions for each dimension was computed, providing a comprehensive measure of each respondent's political efficacy levels.

We pursued the goal of collecting constructive requests for changes to the CPA programme in a final open and non-obligatory question (Question 16): 'What would you like to be changed about the climate pact ambassador mandate?' This open-ended approach allows for detailed qualitative insights into potential improvements. To date, no official evaluations of the CPA programme by the European Commission are publicly available, making our research, though exploratory in nature, valuable for understanding programme effectiveness and providing policy implications.

Data analysis

Given the relatively small size of the sample of CPAs who completed the questionnaire the analyses in the subsequent chapter are based on descriptive statistics. To visualize the results from the survey effectively, we created bar graphs with absolute or relative frequencies, which partly additionally distinguish between key demographic variables, such as age, gender, occupational background, and individual versus organizational mandate.

The data analysis commences with a comparative overview of the total CPA population and our survey sample. We then employ a two-step analytical approach, first examining outcome variables within each of the three dimensions (motivation, strategy, and experience), followed by the integration of explanatory variables as detailed in Table 3.1 in Chapter 3.

Summary

In this chapter, we provided an overview of the data collection methods that align with the specific research interests and objectives, while also addressing the methodological challenges inherent in studying a novel institutional arrangement that combines elements of citizen participation, environmental advocacy, and EU-level policy implementation, making it a fascinating case study to understand contemporary forms of civic engagement.

The ECP website served as an effective and accessible platform for contacting CPAs and inviting them to participate in an online survey. The website's listing of CPAs, along with their contact information, provided an invaluable resource for reaching this specialized population. Furthermore, the ECP Secretariat assisted us in approaching the CPAs. We took the decision to utilize an online survey methodology on the basis of several practical considerations. First, the CPAs are geographically dispersed across multiple EU member states, making face-to-face interviews logistically challenging and financially prohibitive. Second, given that CPAs operate within a digital environment and are accustomed to online communication through their mandate, an online survey represented a familiar and accessible format. Finally, the standardized nature of the survey instrument allows for systematic comparison across different types of CPAs and their experiences.

Despite the institutional support from the ECP Secretariat and the comprehensive outreach efforts, ultimately, only 76 CPAs submitted complete responses, which form the empirical foundation of the findings presented in Chapter 5. While the response rate of 14 per cent was lower than what we initially expected based on the public-facing nature of the CPA role and their presumed engagement with community outreach, the data demonstrated sufficient variation across key variables and, therefore, could be analyzed in a meaningful way and with reasonable confidence in the findings.

As we explained in this chapter, we employed established measurements to operationalize the key variables, drawing systematically from two complementary literature streams: research on the determinants of volunteering and literature on policy advocacy and civic engagement. By utilizing these established measurements, the findings can contribute meaningfully to these broader bodies of literature and can be compared with existing research in these fields.

This methodological approach is valuable because it establishes the groundwork for future research endeavours in this field. Moreover, while the specific format of CPAs may be unique within the European governance landscape, both the drivers for volunteering to become one and the strategies individuals adopt once given the mandate can be generalized more widely to other contexts of civic engagement and environmental activism. Therefore, the systematic compilation of the first CPA database could be one of the key contributions this book makes to the study of the new format of CPAs. This new database supports initial studies with additional empirical insights and establishes the groundwork for future research endeavours in this field.

FIVE

Empirical Findings

Introduction

Having explored how the EU Commission has strategically designed the ECP and its climate ambassador programme, and having developed a comprehensive conceptual framework that identifies age, gender, mandate type, professional background, and political efficacy as key explanatory variables, we turn to the heart of this investigation. This chapter presents the empirical findings and addresses the central research questions about what drives CPAs, how they navigate the policy advocacy landscape, and what their experiences reveal about this ambitious programme.

First, a portrait is painted of the individuals who step forward to become CPAs. Who are these volunteers? Then, we dive into the motivational landscape, exploring the three key dimensions of altruism, egoism, and investment, and uncovering how well the conceptual framework variables (see Chapter 3) explain differences between CPAs. From there, we examine the strategic toolkit that CPAs employ in their policy advocacy work, revealing how individual versus organizational mandates shape their approaches, and discovering the ways professional background influences their chosen strategies.

The analysis then shifts to the lived experiences of CPAs themselves: what has this ambassadorship meant to them? Here, we assess how different motivations and levels of political self-efficacy create different programme experiences. This section offers valuable insights CPAs themselves provide about how the initiative could be strengthened and refined. Then, this chapter concludes with a synthesis of these findings.

Climate Pact Ambassadors population and sample at a glance

Before commencing this analysis, it is worth examining the total CPA population versus the sample respondents of this study. Figure 5.1 shows the distribution of the total CPA population, as shown on the ECP website on 8 October 2023, broken down by the countries where they reside. We chose this date because the data on who is listed on the ECP's website as a CPA is time-sensitive (see Chapter 4) and this corresponds to the month preceding the fielding of the European Climate Pact Ambassador Survey. At that time, the highest number of CPAs, 17.4 per cent, resided in Italy, 16.5 per cent in Spain, and 10.4 per cent lived in Germany, while the lowest numbers could be found in Lithuania (0.4 per cent), Malta (0.4 per cent), and Estonia (0.5 per cent).

Figure 5.2 shows the countries in which the survey respondents reside. From this figure, the participation of CPAs from Italy, Spain, and Germany is proportionate to their representation in the CPA population, which also holds true for several other EU member states, such as Belgium, France, the Netherlands, Poland, and Portugal. Some countries are slightly under-represented, such as Greece, while others are slightly over-represented, most notably Malta. No CPAs participated from the Czech Republic, Estonia, Ireland, Latvia, Lithuania, Luxembourg, or Slovakia. Among this group of countries, the absence of Ireland represents the most severe limitation because it is fairly well represented in the CPA population (see Figure 5.1). These figures also include a residual category, which includes CPAs from non-EU member states, such as Serbia or Ukraine, which at the time of data collection still held their mandate (see Chapter 2).

In terms of professional background, most CPAs in the survey work for non-profit or for-profit organizations (26 per cent

Figure 5.1: CPA population, October 2023

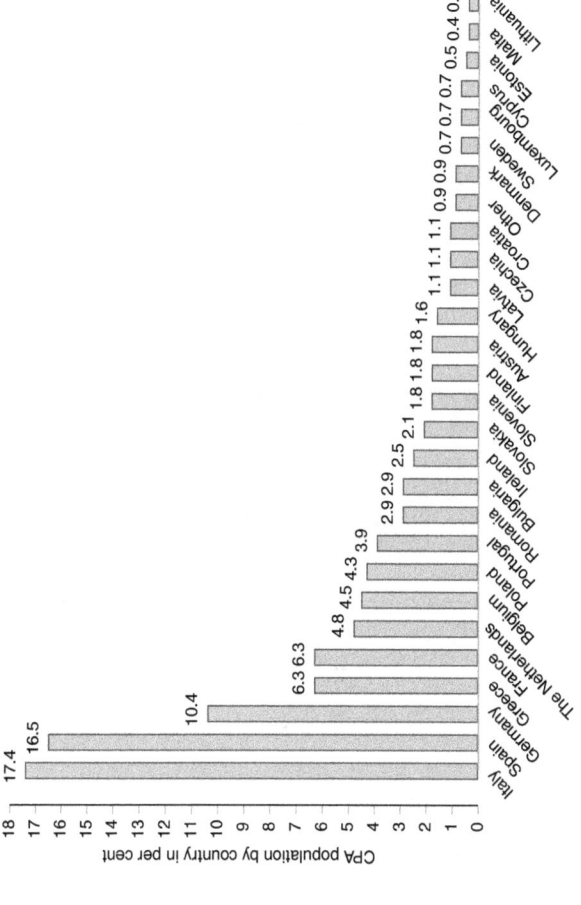

Source: Own data and elaboration

Figure 5.2: Survey respondents, November 2023

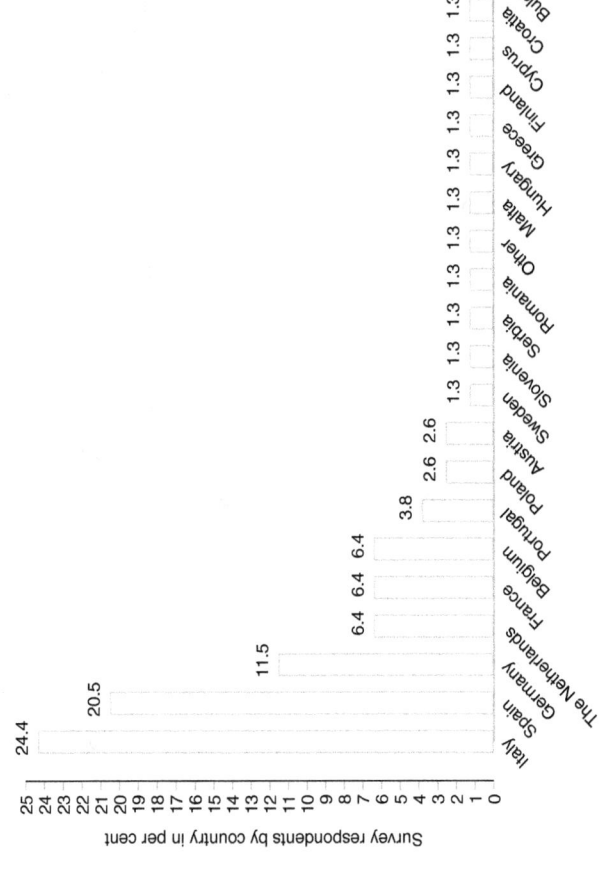

Source: Own data and elaboration

for each category), followed by around 17 per cent of individuals working in academia, 9 per cent have a public sector working background, while 6.4 per cent each work in education or are students.

Motivations

As outlined in Chapter 3, the analysis of motivation draws on Ziemek's (2006) framework, which distinguishes three fundamental types of motivations that drive people to volunteer: altruism (based on genuine concern for the public good), egoism (driven by personal benefits and private consumption), and investment (focused on enhancing one's labour market value through volunteer work). To understand what motivates CPAs, our survey asked respondents to rate their agreement with various motivational statements on a scale from 'strongly disagree' to 'strongly agree'. This approach allowed capturing the nuanced reasons behind their commitment to climate action.

Our survey revealed several distinct, although sometimes overlapping, motivations for joining the programme, as shown in Figure 5.3. These findings paint an interesting picture of why individuals choose to become CPAs.

Starting with altruistic motivational factors, which correspond to the first two responses to the survey question, the results demonstrate a strong moral foundation underlying participation. An overwhelming majority (99 per cent) of CPAs agreed or strongly agreed that social principles and moral obligation drove their engagement, a finding that suggests these individuals are motivated by a desire to contribute to the greater good. Additionally, 59 per cent strongly agreed that there was a significant need for this type of work, indicating that CPAs recognize the urgency of climate action and feel compelled to fill a perceived gap in addressing environmental challenges.

To bring these findings to life and relate the survey questions back to the personal stories that CPAs share on the ECP

Figure 5.3: Motivational drivers for the decision to become a CPA

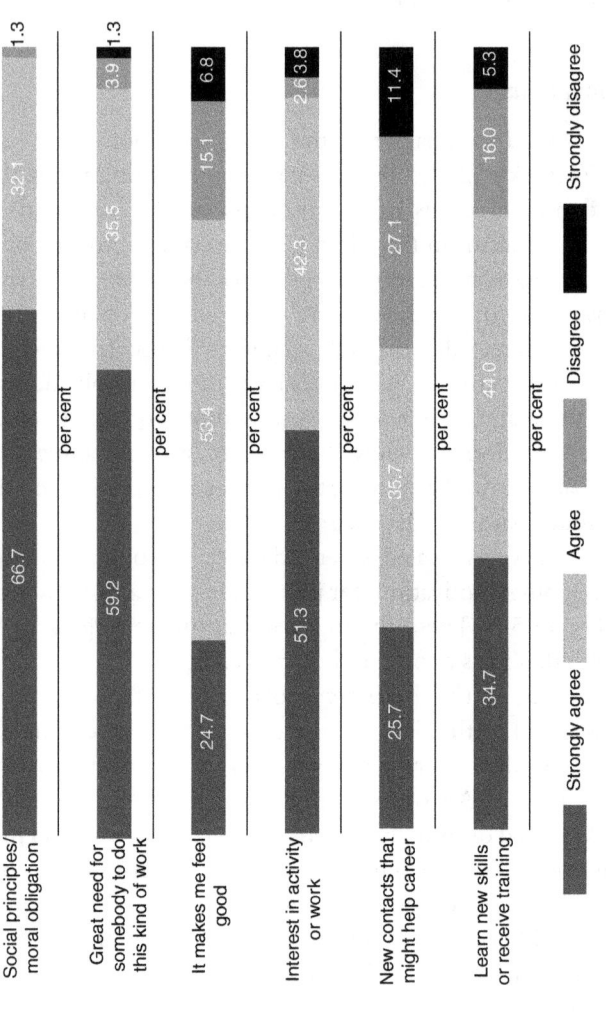

Source: Own data and elaboration

website, we examined individual profiles and testimonials. These personal accounts provide support for the altruistic motivations revealed in the survey data.[1] As one CPA from Greece eloquently stated on the website: 'In my pursuit as an alpinist, to climb the highest summit of each of the seven continents, I travelled to the most remote places on Earth, witnessed the degradation of our planet's natural environment and decided to act' (European Commission, 2025c). This testimony illustrates how direct experience with environmental degradation can transform personal concern into concrete action. Similarly, another CPA from Hungary described her commitment in her profile: 'Whether it is nature, biodiversity, sustainable production, green solutions, responsible consumption or a change of attitude – I will be there to help!' (European Commission, 2025c). This statement reflects the comprehensive approach many CPAs take, recognizing that climate action requires engagement across multiple domains.

Responses indicating that individuals feel good serving as CPA and that they take a general interest in climate action correspond to the second set of motivational drivers, capturing egoism, that is, motivations driven by personal satisfaction and individual interests rather than altruistic concerns. It is important to note that 'egoistic' motivations in this context are not negative; they simply reflect the personal benefits and satisfaction that individuals derive from their volunteer work (see Chapter 3).

The responses to these questions revealed interesting and somewhat mixed patterns. Regarding emotional reward, opinions were notably divided. While approximately one-quarter of respondents cited the emotional satisfaction of serving as a CPA as a strong driver for their application to the programme, 22 per cent rejected it as an influential factor in pursuing a mandate. This split suggests that while some CPAs are motivated by the personal fulfilment that comes from climate action, others are driven by factors that extend beyond individual emotional gratification.

In contrast, the data revealed overwhelming consensus on another egoistic motivation: a remarkable 94 per cent of respondents (strongly) agreed that they are acting as CPA because they take a general interest in climate action. This finding indicates that personal interest and curiosity about climate issues serve as powerful motivational forces.

This motivation is illustrated by a CPA from Ireland, who provides a particularly compelling example of how personal interest intertwines with broader concerns. In his profile, he states: 'It has been thirty years since I was first elected as a public representative. I am conscious that together on this planet, we do not have another thirty years in which to get things right. More selfishly I want a planet that my grandchildren can live on sustainably' (European Commission, 2025c). This candid reflection demonstrates how personal stakes, in this case, concern for future generations, can serve as a driving force for climate action, even when the individual acknowledges the 'selfish' nature of wanting to ensure a liveable planet for one's own family.

Developing new contacts that help advance one's career and learning new skills or benefitting from training opportunities represent the final two response categories shown in Figure 5.3, capturing investment-driven motivations. These motivations reflect the strategic aspect of volunteering, where individuals seek to enhance their professional prospects and capabilities through their engagement as CPAs.

Regarding networking opportunities, the survey results revealed a clear pattern: approximately 39 per cent of respondents (strongly) disagreed with the statement that they are serving as CPA in order to develop new contacts that could advance their careers. This suggests that the majority of CPAs are not primarily motivated by professional networking considerations. This empirical picture is supported by the CPAs' public profiles, because only a few explicitly refer to such networking motivations in their personal statements.

Nonetheless, there are some notable examples that demonstrate how professional development and climate action can intersect meaningfully. These cases, drawn directly from the public web profiles, illustrate that while networking may not be the primary driver for most CPAs, it can still play a complementary role in their motivation. A case in point is this statement by one Spanish CPA: 'Wishing to expand contacts with groups and entities related as professional in the marketing sector and involved with the business fabric and economic development in the Canary Islands, I feel committed to a better sustainable evolution for our environment' (European Commission, 2025c). This statement demonstrates how professional networking aspirations can align with genuine environmental commitment, creating a synergy between career development and – in this case - environmental action.

The second indicator of investment-driven motivation, skill development and training opportunities, resonated much more strongly with survey respondents. Approximately 79 per cent of participants acknowledged that learning new skills or benefitting from training played a role in their decision to become CPAs. This finding suggests that while social relationships and networking may not be a primary motivator, the opportunity for personal and professional development through skill acquisition is highly valued by CPAs.

Overall, the findings reveal a clear motivational hierarchy among CPAs. Both altruism and egoism emerged as the primary driving forces, while investment considerations played a considerably less important role in respondents' initial decisions to join the programme. This pattern suggests that CPAs are fundamentally motivated by a combination of genuine concern for the public good and personal interest in climate issues, rather than by strategic career considerations.

However, when examining egoistic motivations more closely, the data revealed an important distinction. The respondents perceived general interest in working on the issue of climate change as significantly more relevant than the emotional reward they could obtain from this activity. This finding indicates that while CPAs do derive personal satisfaction from their work, their engagement is driven more by curiosity for the subject matter than by the emotional gratification of volunteering itself.

Interestingly, despite this study focusing on a very specialized type of volunteering and drawing from a distinct sample of climate-/environment-focused individuals, the findings shown in Figure 5.3 align remarkably well with those established by Ziemek (2006) in her research. Her findings, based on representative samples of volunteers across diverse contexts in Bangladesh, Ghana, Poland, and South Korea, revealed that altruism and egoism were approximately equally prevalent as motivating factors, while investment-related motivations were marginally less significant. This convergence in empirical findings suggests that the motivational patterns observed among CPAs may reflect broader, universal principles of volunteer engagement, transcending both cultural boundaries and the specific domain of climate and environmental action.

As shown in Figure 5.4, a high share of CPAs working in education, the non-profit sector, the private sector, as well as academia, cited altruistic motivations, such as social principles and the need for somebody to do this kind of work, as influential for their decision to volunteer. For egoistic motives, such as emotional reward or the interest in the activity, the empirical picture is more mixed. Those working in education only agreed that emotional rewards are important, whereas students and people from the private, public, and non-profit sectors had a higher disagreement rate. With the exception of people working in the non-profit sector, all other occupational status groups indicated that they take an interest in the activity, that is, in acting as CPAs.

Concerning the motivation based on investment considerations, students indicated that both extending their networks and learning new skills is important to them. In contrast, most people from the public sector did not cite their motivation to become a CPA with establishing new contacts for careers, and workers in the education sector did

EMPIRICAL FINDINGS

Figure 5.4: Motivational factors by professional background

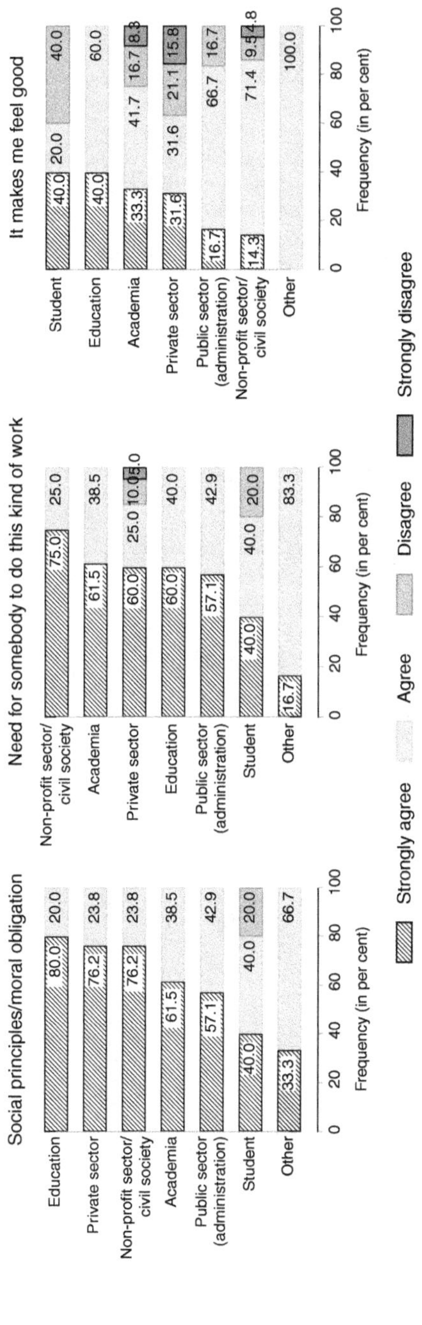

Social principles/moral obligation

- Education: 80.0 | 20.0
- Private sector: 76.2 | 23.8
- Non-profit sector/civil society: 76.2 | 23.8
- Academia: 61.5 | 38.5
- Public sector (administration): 57.1 | 42.9
- Student: 40.0 | 40.0 | 20.0
- Other: 33.3 | 66.7

Frequency (in per cent)

Need for somebody to do this kind of work

- Non-profit sector/civil society: 75.0 | 25.0
- Academia: 61.5 | 38.5
- Private sector: 60.0 | 25.0 | 10.0 | 5.0
- Education: 60.0 | 40.0
- Public sector (administration): 57.1 | 42.9
- Student: 40.0 | 40.0 | 20.0
- Other: 16.7 | 83.3

Frequency (in per cent)

It makes me feel good

- Student: 40.0 | 20.0 | 40.0
- Education: 40.0 | 60.0
- Academia: 33.3 | 41.7 | 16.7 | 8.3
- Private sector: 31.6 | 31.6 | 21.1 | 15.8
- Public sector (administration): 16.7 | 66.7 | 16.7
- Non-profit sector/civil society: 14.3 | 71.4 | 9.5 | 4.8
- Other: 100.0

Frequency (in per cent)

▨ Strongly agree ▤ Agree ▨ Disagree ■ Strongly disagree

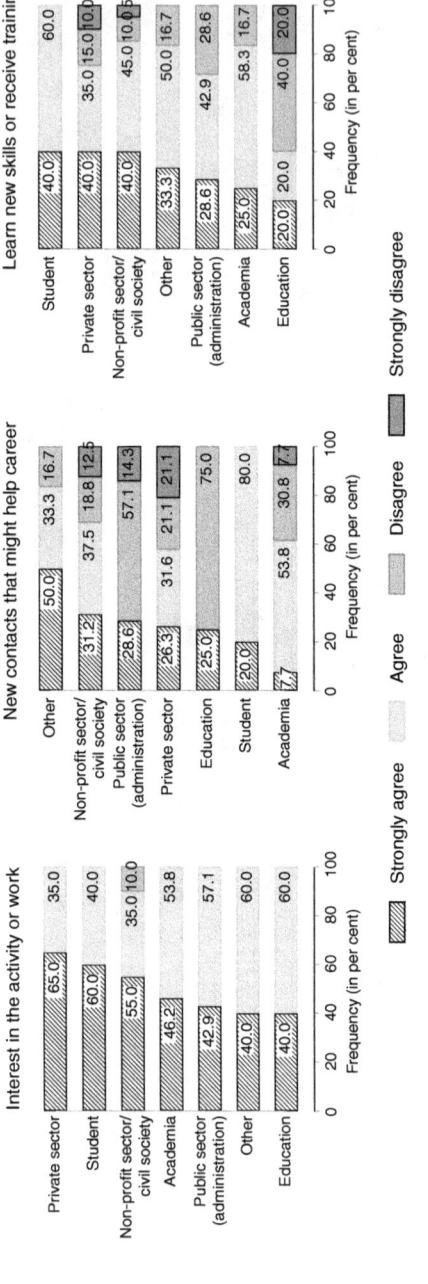

Figure 5.4: Motivational factors by professional background (continued)

Source: Own data and elaboration

not state that to learn new skills or receiving training was the main motivation to act as CPA. This shows that occupational background is an important explanatory factor when it comes to explaining variation in the three types of motivations indicated by the CPAs.

While the previous question allowed respondents to select multiple answers, which captures the full spectrum of motivations that influence their participation, the data shown in Figure 5.5 is based on responses to a follow-up question that asked participants to identify their single most important motivator. This 'forced-choice' approach represents a deliberate methodological decision designed to reveal respondents' core priorities when compelled to resolve trade-offs between different motivational factors. The answer to this single-choice question was used to assess which factors, in addition to professional background, may explain variation between CPAs. As explained in Chapter 4, the corresponding survey question is based on a compilation of items from Ziemek (2006) and the Australian Bureau of Statistics (2021). It contains a slightly different wording to cross-validate the findings shown in Figure 5.3 and Figure 5.4, which were solely based on Ziemek (2006).

The data revealed the perceived societal need for committed individuals as the primary motivating factor, with 32 respondents citing it as their most important reason for becoming a CPA. This finding strongly resonates with the importance of altruism as already shown by the responses to the item 'great need for somebody to do this kind of work' in Figure 5.3, reinforcing the conclusion that CPAs are fundamentally driven by a sense of moral obligation and recognition of the urgency of climate and environmental challenges.

To move beyond mere data description and offer deeper insights into the relationships underlying these motivational patterns, Figure 5.5 shows the responses by key demographic and structural variables: gender, age, and mandate type. This disaggregated analysis allowed the examination of whether

EUROPEAN CLIMATE PACT AMBASSADORS

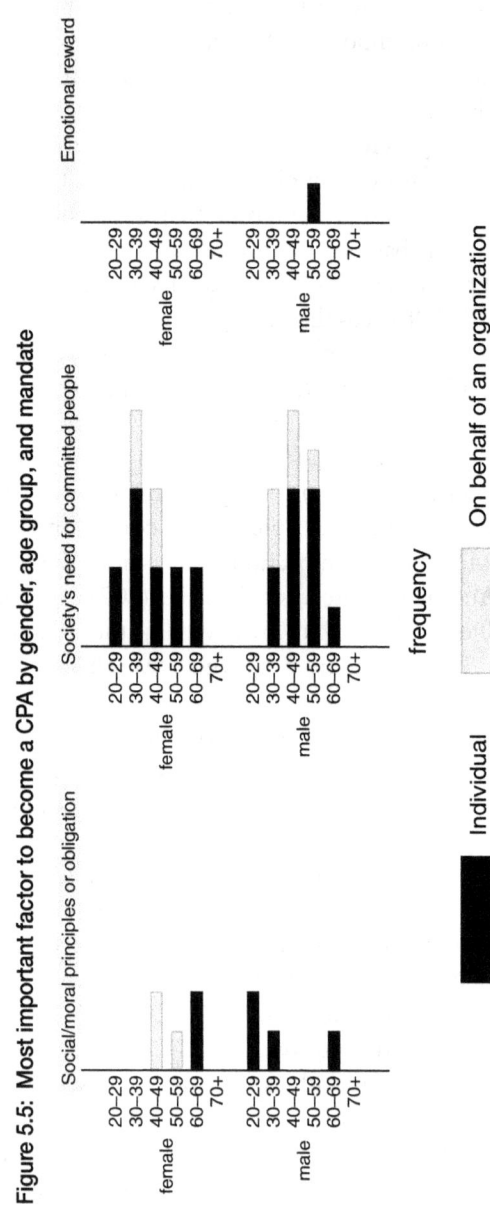

Figure 5.5: Most important factor to become a CPA by gender, age group, and mandate

EMPIRICAL FINDINGS

Figure 5.5: Most important factor to become a CPA by gender, age group, and mandate (continued)

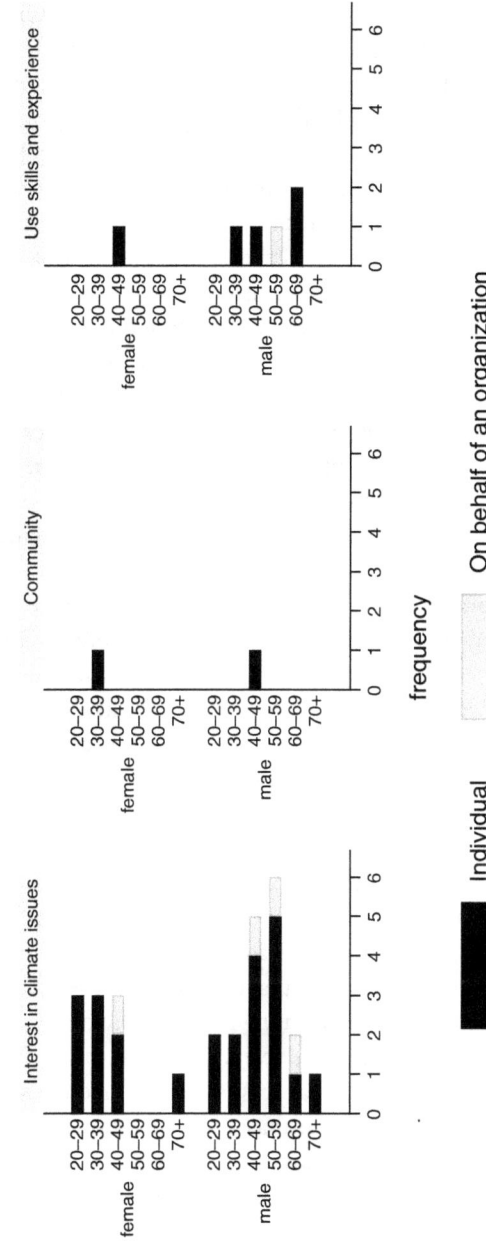

Note: Results for non-binary category: two respondents with individual mandate named 'Society's need for committed people' as the main decisive factor, one belonging to the age group 20–29 years and one among the 30–39-year-olds. The dimensions 'Social/moral principles or obligation' and 'society's need for committed people' correspond to altruism, 'emotional reward' and 'interest in climate issues' correspond to egoism, and 'community' and 'use skills and experience' capture investment-based motivations.

Source: Own data and elaboration

motivational priorities vary across different groups of CPAs, potentially revealing important differences in how various segments of the CPA community approach their mandate.

This detailed analysis yielded a striking result. Across all demographic categories – gender, age, and mandate type (individual vs. organizational) – respondents demonstrated remarkably consistent views on the importance of perceived societal needs, which emerged as the dominant motivational factor (see Figure 5.3). This consistency suggests that the recognition of urgent societal need for climate action transcends traditional demographic boundaries and organizational contexts, pointing to a shared understanding among CPAs about the critical nature of their work. The uniformity of this response across different groups strengthens the argument that altruistic motivation, particularly the perception of societal need, represents a fundamental driver of engagement within the CPA community.

Social or moral principles as altruistic motives were less frequently cited as important overall, though eight CPAs identified them as their primary motivation – with female organizational representatives being particularly likely to cite these principles as their main reason for volunteering. In particular, the findings for the sense of obligation are worth flagging since they concur with the results by Tosun et al (2023a). In their analysis of the short texts published alongside the CPAs' profiles on the ECP website, the authors also found that very few refer to moral obligations, more precisely to current and future generations, in order to motivate their participation in the programme. The authors showed that organization-based CPAs are less likely to state moral obligation vis-à-vis individual CPAs.

Thus, while the data sources are very different, the findings are similar. It is conceivable that this is a difference between those volunteering for the ambassador programme and, therefore, pursuing the path of advocacy (see Parks et al, 2023), and those who pursue climate activism; the latter could refer more to the social obligation to act as climate activists.

Interest in climate issues, as an egoistic motivational driver, emerged as the second most frequently cited factor overall, indicated by 28 respondents. This response, too, aligns with the previous findings shown in Figure 5.3, where the majority of respondents stated an interest in their activity as a CPA, stressing the importance of egoistic motivations for volunteering as a CPA. This motivation was particularly prevalent among individual mandate holders, men in their forties and fifties, and women in their twenties and thirties. In other words, the response patterns by age and gender are less consistent for interest in climate issues vis-à-vis the perceived need of society for climate action.

The utilization of existing skills and experiences as an investment-driven factor was notably important among male respondents aged 30 and above. This is certainly an interesting finding and somewhat unexpected, which warrants further investigation by future research on this topic.

Other factors, including emotional reward and community, were cited as primary motivations by only a small number of respondents. Therefore, we did not consider it meaningful to connect them with any potential explanatory variable.

To summarize the findings obtained so far, the analysis revealed diverse motivations for becoming a CPA, with some variation across gender, age, mandate type, and occupation. Gender differences were most pronounced in skill-related motivations, which correspond to investment-driven motivations, with male respondents more frequently citing both the application of existing skills and the acquisition of new ones as motivating factors. Organizational mandate holders predominantly emphasized their desire for societal impact and social principles as their motivations. Networking and social contact development were relatively minor motivational factors, which is unsurprising given that CPA applicants are required to be leaders of formal or informal communities, suggesting pre-existing strong networks.

Furthermore, as shown by Figure 5.4, professional background mattered for all three outcomes of motivational reasons to volunteer as a CPA, showing the greatest variation among different professional groups for investment-based motivations.

Strategies

As explained in Chapter 2, the ambassador programme stimulates a wide range of ambassador-led activities, including different types of policy advocacy. These activities demonstrate the diverse approaches that CPAs adopt in their efforts to promote climate change and environmental awareness and stimulate action within their respective communities and beyond. The programme's decentralized structure allows ambassadors autonomy in designing and implementing initiatives that align with both the ECP's overarching objectives and their local contexts and expertise.

Before turning to the assessment of the advocacy-related activities, it is instructive to engage with some examples in order to gain a better understanding of the real world implementation of the ambassadorship. These illustrative cases reveal the multifaceted nature of ambassador engagement, ranging from grassroots community mobilization and educational initiatives to more formal policy intervention and institutional collaboration.

A CPA based in Italy organized a climate walk around Venice in order to explore the relationship between the city of Venice and the lagoon and to discuss the effects of climate change with the participants.[2] A CPA based in Denmark helped to set up a sustainable farming stand at an annual 3-day climate event called *Klimafolkemødet* (The Climate People's Meeting). The CPA utilized the large visitor numbers to the event to engage in debates and raise awareness about the ECP and the importance of sustainable farming.[3] A CPA based in the Netherlands collaborated with members of his community to install solar panels on the roofs of houses and two primary schools.[4] Several CPAs based in Greece participated in television broadcast formats in which they elaborated on the circular economy and just transition.[5] A CPA based in Spain has launched a crowdfunding campaign to install solar panels and heat pumps at a school.[6] A CPA in Bulgaria hosted a Peer Parliament[7] at the International Fair in Plovdiv, where she discussed mobility, energy, and consumption with over 40 students aged 12–18 from that area, along with their teachers.[8]

Among the activities carried out are advocacy actions. For example, a group of CPAs based in Sweden published a call on the social networking platform LinkedIn addressing the members of the European Parliament and asking them to maintain the level of ambition of policies on climate change, to increase climate preparedness, while equally calling on them to protect democracy and support civil society.[9] A CPA based in the Slovak Republic co-organized a petition entitled 'For the Climate.' It obtained sufficient signatures to require the Slovak parliament to discuss the climate.[10] Furthermore, the ECP flagship events offer CPAs the opportunity to engage with Wopke Hoekstra, the EU Commissioner responsible for the ECP and climate action in the EU.[11]

To commence the analysis, Figure 5.6 shows the absolute frequency of CPAs engaging in different types of advocacy activities. The data reveals a clear pattern in advocacy behaviour, with most CPAs adopting a diverse approach to their advocacy efforts. Most respondents, 68 in total, reported having worked in another organization. Posting about politics online and signing petitions, as well as seeking access to politicians as an advocacy act, also ranked as favoured strategies amongst the majority of CPAs. These activities represent relatively accessible forms of political engagement that can be pursued with minimal barriers to entry. In contrast, wearing campaign badges, working within a political party, and participating in demonstrations proved less appealing to the majority of surveyed CPAs. To explain the varying patterns when pursuing strategies, in the following section we analyze the influence of mandate type and the professional background of CPAs on their advocacy activities.

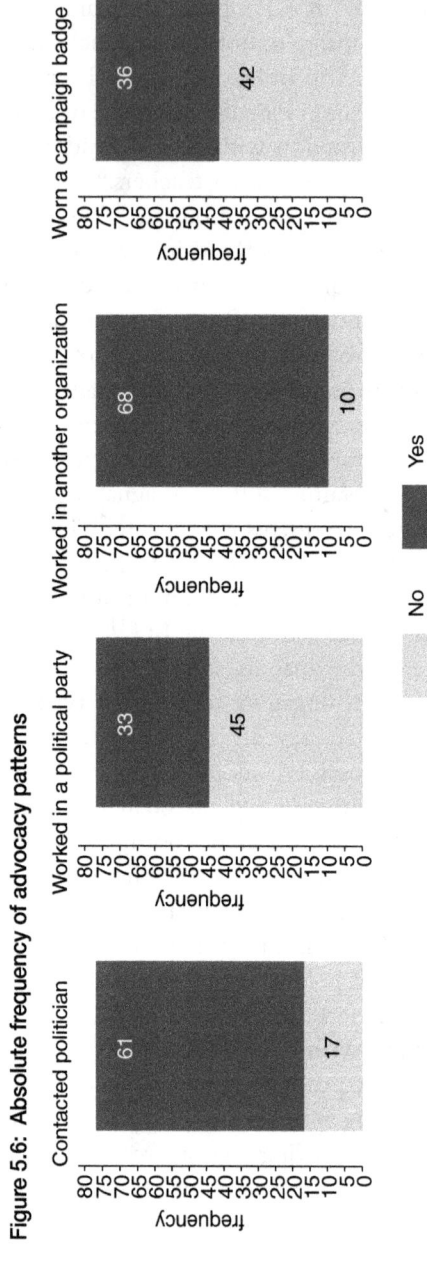

Figure 5.6: Absolute frequency of advocacy patterns

Figure 5.6: Absolute frequency of advocacy patterns (continued)

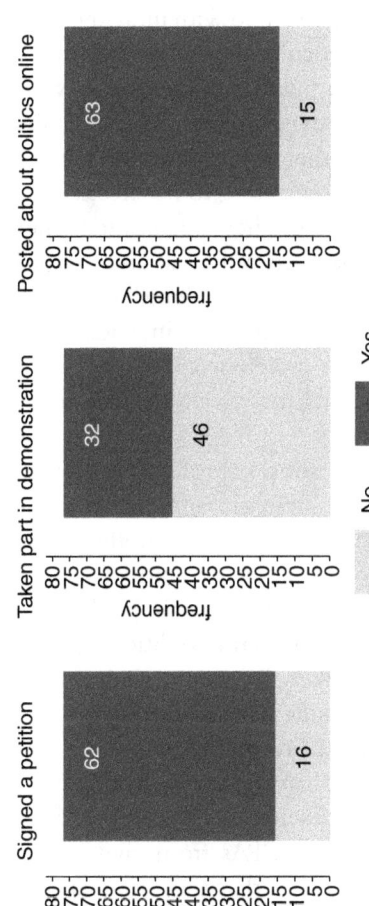

Source: Own data and elaboration

Figure 5.7 shows distinct strategic differences in the advocacy approaches between individual CPAs and those acting on behalf of organizations. Across all activity types, a higher percentage of CPAs with organizational mandates pursued policy advocacy compared with individual CPAs. This finding is reasonable, particularly given that organizational members tend to represent public bodies, NGOs working on climate change and the environment, and companies offering green technologies or other types of business solutions for mitigating climate change or adapting to its effects.

Working within another organization emerged as the most preferred approach to policy advocacy for both mandates, with 94 per cent of respondents with organizational CPA mandates and 86 per cent with individual mandates reporting this activity. This preference likely reflects the institutional resources and established networks that organizations provide for advocacy work.

As Figure 5.8 shows, professional background notably influences CPAs' strategic approaches to advocacy, except for the activity of working in another organization, which was, as mentioned earlier, the most favoured activity among all groups. CPAs from academia predominantly focused on signing petitions, contacting politicians, and sharing political content online, likely leveraging their research expertise and communication skills. Those working in education prioritized posting about politics online and signing petitions, as well as contacting politicians or working in political parties, suggesting a preference for visible yet moderate forms of political expression. CPAs from civil society and the non-profit sector indicated that they are primarily engaged with contacting politicians, posting about politics, and taking part in demonstrations.

CPAs from the private sector showed diverse engagement patterns, with relatively high participation rates in signing petitions, contacting politicians, and posting about politics online.

EMPIRICAL FINDINGS

Figure 5.7: Different ways of pursuing climate action by mandate since the beginning of CPA term

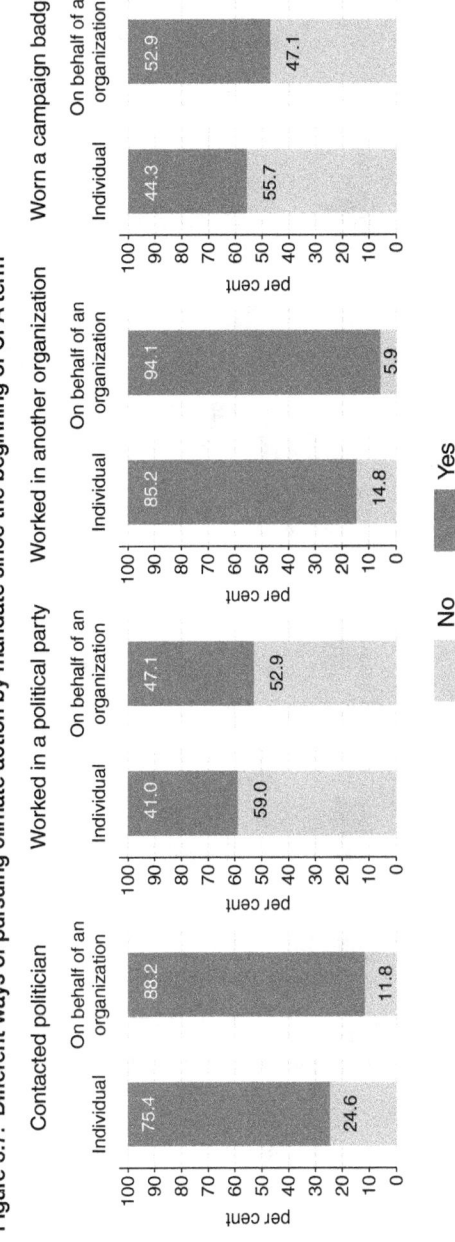

(continued)

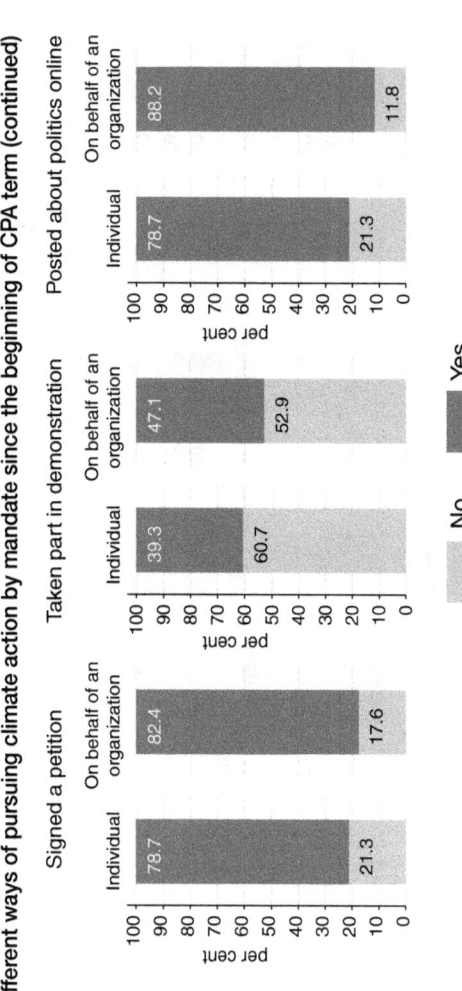

Figure 5.7: Different ways of pursuing climate action by mandate since the beginning of CPA term (continued)

Source: Own data and elaboration

EMPIRICAL FINDINGS

Figure 5.8: Taking on climate action by occupational status

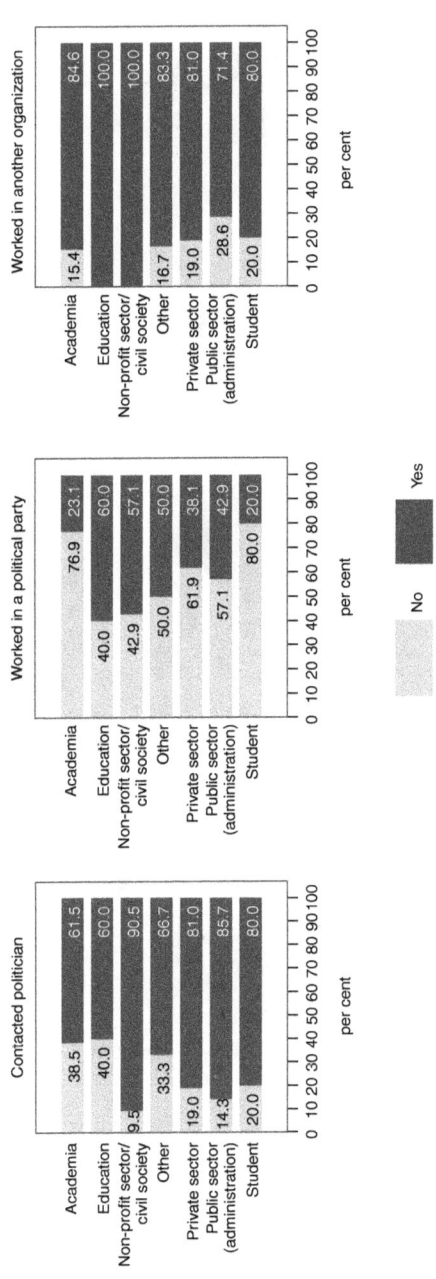

(continued)

EUROPEAN CLIMATE PACT AMBASSADORS

Figure 5.8: Taking on climate action by occupational status (continued)

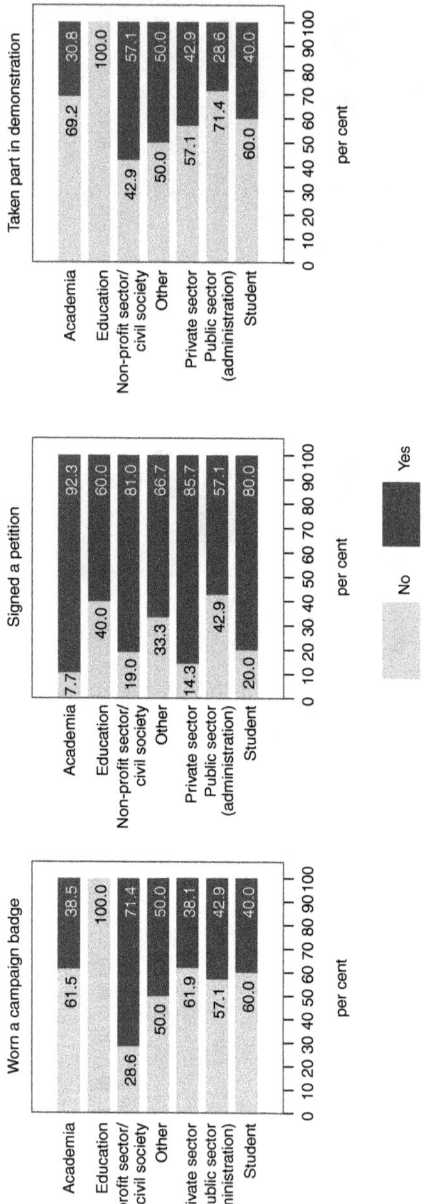

Worn a campaign badge

Occupation	No	Yes
Academia	61.5	38.5
Education		100.0
Non-profit sector/civil society	28.6	71.4
Other	50.0	50.0
Private sector	61.9	38.1
Public sector (administration)	57.1	42.9
Student	60.0	40.0

per cent

Signed a petition

Occupation	No	Yes
Academia	7.7	92.3
Education	40.0	60.0
Non-profit sector/civil society	19.0	81.0
Other	33.3	66.7
Private sector	14.3	85.7
Public sector (administration)	42.9	57.1
Student	20.0	80.0

per cent

Taken part in demonstration

Occupation	No	Yes
Academia	69.2	30.8
Education		100.0
Non-profit sector/civil society	42.9	57.1
Other	50.0	50.0
Private sector	57.1	42.9
Public sector (administration)	71.4	28.6
Student	60.0	40.0

per cent

EMPIRICAL FINDINGS

Figure 5.8: Taking on climate action by occupational status (continued)

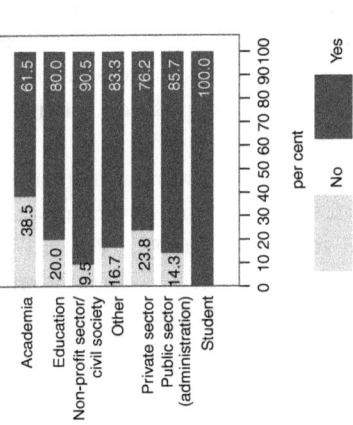

Source: Own data and elaboration

This breadth of activities may reflect the varied corporate cultures and individual motivations within the private sector. In contrast, public sector CPAs mainly limited their engagement to contacting politicians and posting about politics online and have the lowest share in taking part in a demonstration, possibly due to professional constraints or codes of conduct that restrict more overt political activities while in public employment. Students showed a distinct preference for posting about politics online, signing a petition, or contacting politicians.

Turning now to the CPAs' access to different political stakeholders, Figure 5.9 shows which advocacy strategies CPAs applied, given that they either had an individual or an organizational mandate. A higher share of CPAs representing an organization contacted ministers or elected Members of Parliament (MPs) from opposition parties compared with those CPAs with individual mandates. In each group, CPAs with individual and organizational mandates, more than the majority, contacted MPs from governing parties or majority parties. In parliamentary systems, as most EU member states have in place, MPs affiliated with the majority parties are more influential in shaping climate and environmental policy. Thus, the advocacy pattern indicates that CPAs prioritize accessing influential politicians who can impact agenda-setting and policy making, rather than pursuing judicial oversight or engaging with opposition voices who may have limited legislative power.

Courts seemed less relevant for both CPA groups, even though at least 18 per cent of those from an organization indicated having accessed the courts compared with only 8 per cent of individual CPAs. This approach contrasts with recent trends where individuals have increasingly engaged courts to enforce climate laws (Weller and Tran, 2022). For example, six young Portuguese citizens filed a case against 32 European states in the European Court of Human Rights to align policies with Paris Agreement goals, which proved unsuccessful, and Swiss senior women achieved success in a similar case.

EMPIRICAL FINDINGS

Figure 5.9: Access to political stakeholders by mandate

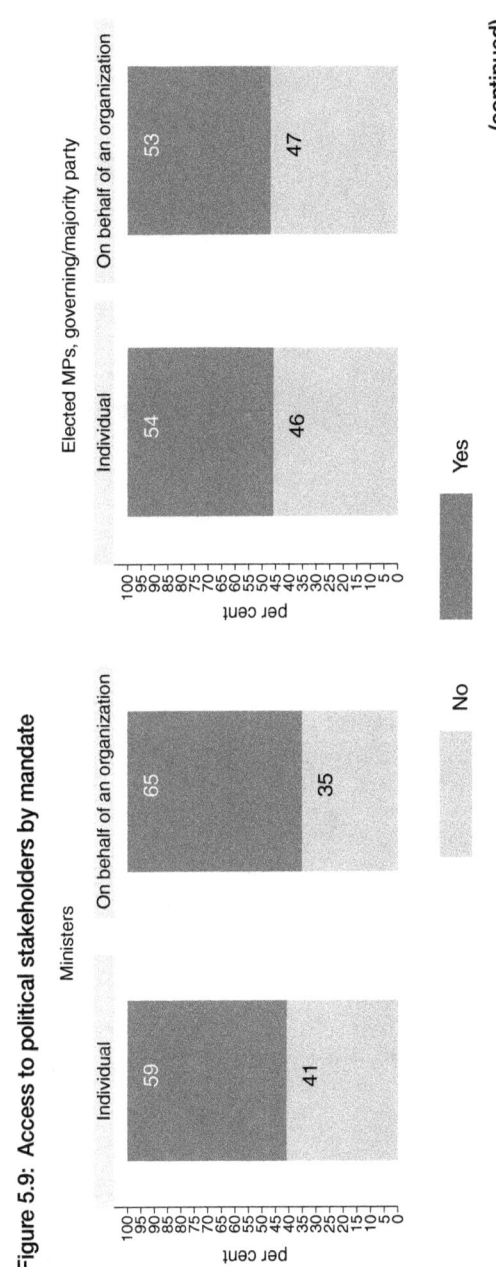

(continued)

EUROPEAN CLIMATE PACT AMBASSADORS

Figure 5.9: Access to political stakeholders by mandate (continued)

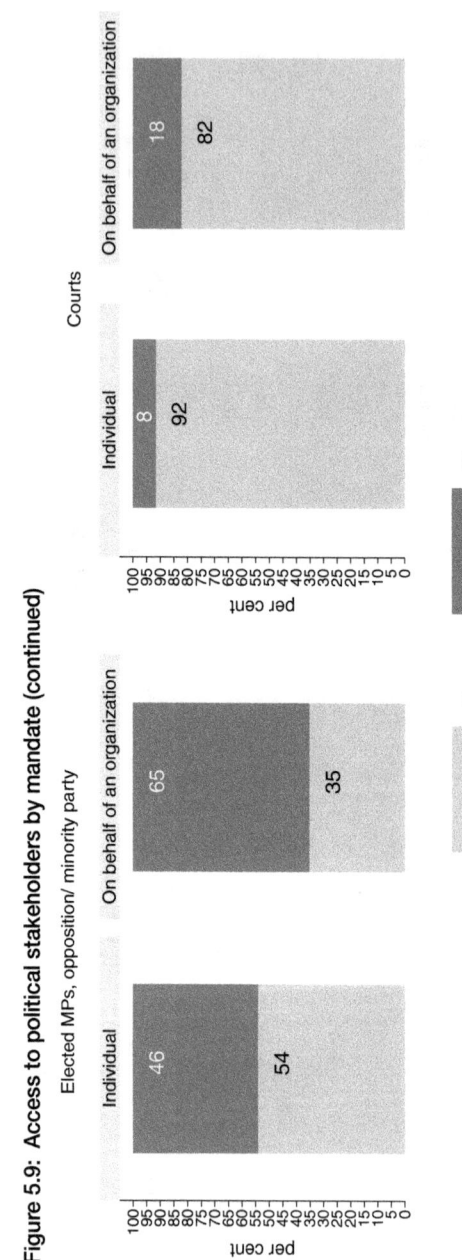

Source: Own data and elaboration

However, the significant bureaucratic hurdles and time-intensive nature of legal proceedings may explain why CPAs generally avoid pursuing judicial action, especially in light of the relatively short duration of their mandate.

Finally, how the CPAs' occupational background influences their way of pursuing access to important political actors was studied. Figure 5.10 shows that people working in the NGO sector or civil society had the highest share of pursuing access to ministers, and elected MPs from the governing as well as opposition parties. This could be due to the fact that they are well-connected and have easier access to institutions in general through events and other forms of organizational meetings. CPAs from the private sector, on the other hand, placed priority on getting access to the high-level political stakeholders, such as ministers and MPs from parties in government. CPAs working in education mainly approached ministers, who are responsible for influencing the direction of education policy on climate change and environmental issues. Fewer than the majority of CPAs with an academic occupational background sought access to ministers, even fewer to elected MPs from either party. CPAs affiliated with the public sector mainly focused on ministers, which is less surprising given that some might be involved in implementing policies of the ministries. This group also had the highest proportion of CPAs who engaged with courts, likely because legal training is common in public sector positions and provides a deeper understanding of court rulings and judicial processes.

In summary, policy advocacy represents an important and multifaceted activity of CPAs, substantiating our view that their mandate extends considerably beyond the simple provision of climate and environmental information. The ambassadors use a comprehensive portfolio of strategies to advance their policy advocacy efforts. Online activism, including posting about politics online or signing petitions, emerged as a particularly popular form of policy advocacy among CPAs. These digital strategies offer CPAs effective means of reaching broader

EUROPEAN CLIMATE PACT AMBASSADORS

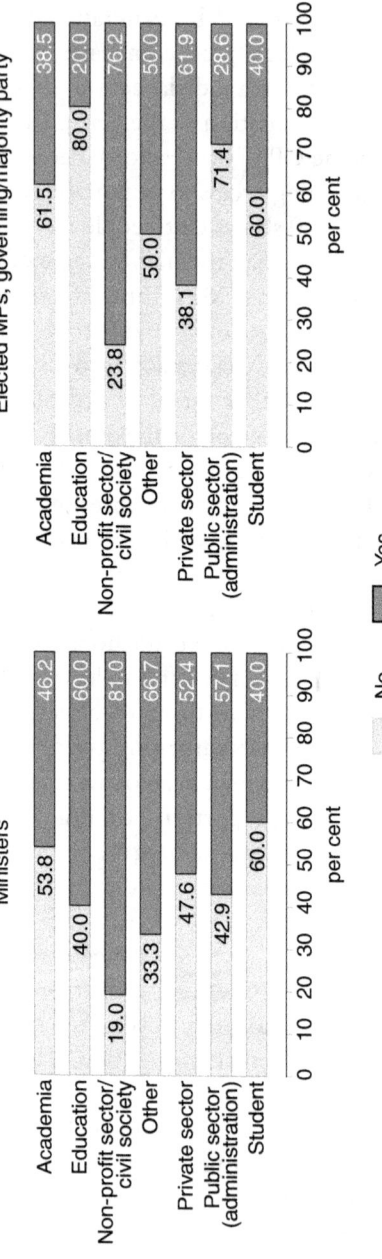

Figure 5.10: Access to political stakeholders by occupation

Figure 5.10: Access to political stakeholders by occupation (continued)

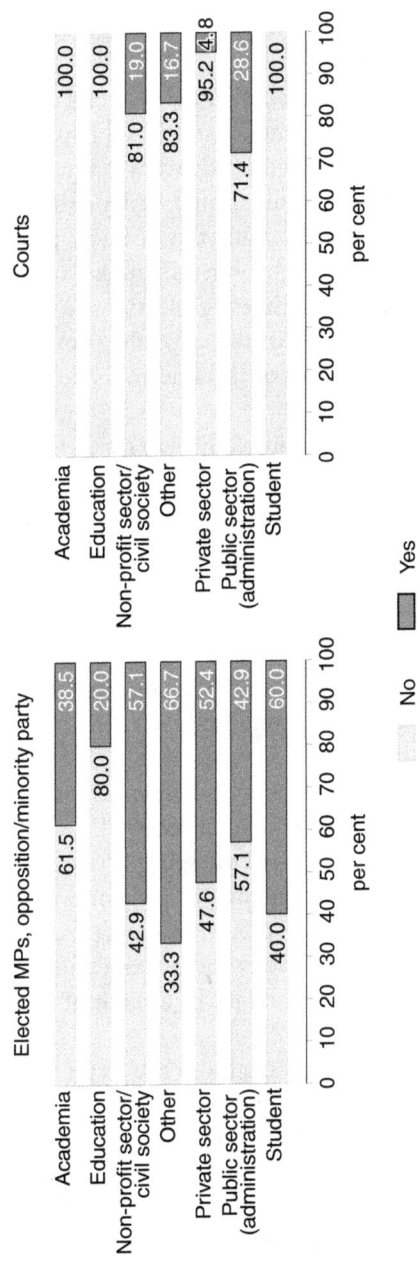

Source: Own data and elaboration

audiences and amplifying their message beyond traditional political channels.

Less frequent activities included working in political parties or participating in demonstrations. This preference pattern suggests that CPAs favour advocacy approaches that allow them to maintain professional independence while still engaging meaningfully with the political process. The vast majority of CPAs, both with individual and organizational mandates, worked in other organizations alongside their CPA role, indicating that their advocacy efforts are often embedded within existing institutional frameworks rather than pursued in isolation.

Both groups showed similarly high rates of engagement with politicians during their terms, primarily targeting ministers and MPs from majority or governing parties, that is, those parties that are particularly influential in policy-making processes. This strategic focus on decision makers with actual legislative power reflects a pragmatic approach to advocacy that prioritizes potential impact over symbolic gestures. Courts remained notably absent from most CPAs' direct advocacy strategies, despite the growing prominence of climate litigation in the broader environmental movement.

Professional background was influential for access to political stakeholders and showed that especially CPAs working in NGOs or civil society had the greatest share of access, which can be explained through their existing networks and potential to put pressure on the policy-making process. Those CPAs with public sector employment were more likely to gain access to courts, which might be a result of their training and expertise in law.

To sum up, both variables, the type of mandate and professional background, proved useful for explaining variation across CPAs. The type of mandate is particularly important for explaining the prevalence of advocacy activities, whereas the professional background can better explain CPAs' strategic approaches to different types of advocacy.

Experience

In a third and final step, we assessed the experience CPAs have gained with their mandate. As shown in Figure 5.11, approximately 62 per cent of the CPAs who participated in the survey indicated that their experience with the ambassador format was (very) positive, followed by about 26 per cent stating that their experience was mixed, and only approximately 13 per cent stating that their experience was (very) negative.

The high share of respondents stating that their experience was (very) positive is likely to capture both the genuine impressions of the CPAs and the potential influence of a

Figure 5.11: CPAs' experience with the mandate

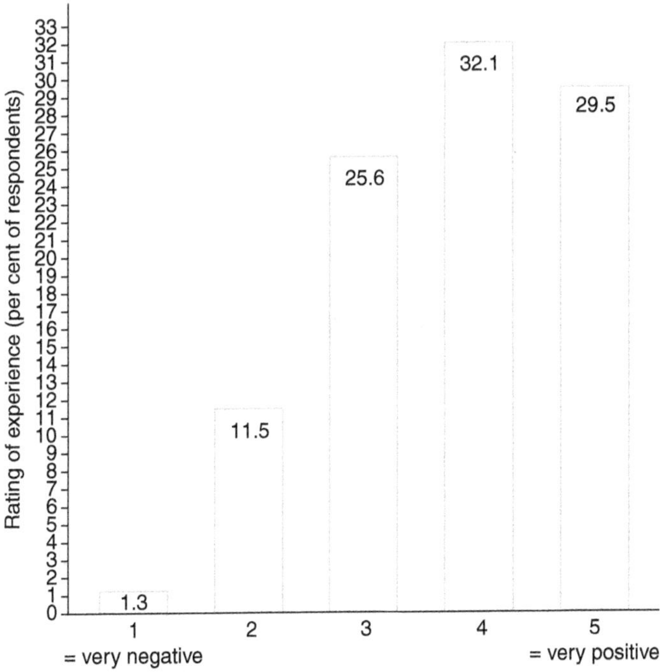

Source: Own data and elaboration

positivity bias due to the self-selection of participants into the survey. Those with more favourable experiences may have been more inclined to participate in the research, potentially skewing the results towards more positive assessments. Despite this potential bias, approximately half of the survey respondents provided different assessments, ranging from mixed to negative experiences. This variation underlines previous observations about the variability in the data and indicates that its quality is sufficiently robust for an exploratory analysis.

The diversity of experiences reported suggests that the ambassador format does not uniformly satisfy all participants, highlighting the importance of understanding the factors that contribute to both positive and negative experiences within the programme. Such insights could inform targeted improvements to enhance participant satisfaction and program effectiveness.

Variations in the motivational dimensions of altruism, egoism, and investment might reflect back on the personal perceptions of the programme, which is why we included them as explanatory factors at this point, as shown in Figure 5.12. In general, those who agreed with the motivational statements were also more often satisfied with the programme. In contrast, those who disagreed with the motivational drivers tended to rate their experience as negative. Hence, motivation per se seems to influence the rating of the ECP ambassador programme, suggesting a clear relationship between initial expectations and subsequent satisfaction.

More than 60 per cent of CPAs with altruistic motives, who (strongly) agreed that moral obligation and the need for this work was their incentive to volunteer, rated their mandate as very positive or positive. Interestingly, for those who disagreed with these altruistic motivations, they still considered their experience positive, indicating that satisfaction can transcend initial motivational frameworks. For the second altruistic indicator, it was reversed, disagreement rather related to dissatisfaction with the mandate.

EMPIRICAL FINDINGS

Figure 5.12: Rating of experience by motivational drivers

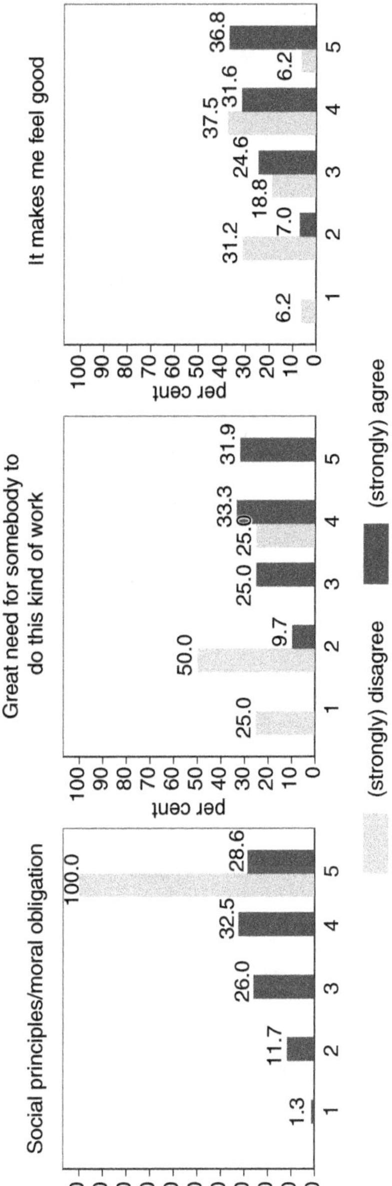

Social principles/moral obligation
- 1: 1.3
- 2: 11.7
- 3: 26.0
- 4: 32.5 / 100.0
- 5: 28.6

Great need for somebody to do this kind of work
- 1: 25.0 / 50.0
- 2: 9.7
- 3: 25.0 / 25.0
- 4: 33.3
- 5: 31.9

It makes me feel good
- 1: 6.2
- 2: 31.2 / 7.0
- 3: 18.8 / 24.6
- 4: 37.5 / 31.6
- 5: 36.8 / 6.2

(strongly) disagree (strongly) agree

(continued)

EUROPEAN CLIMATE PACT AMBASSADORS

Figure 5.12: Rating of experience by motivational drivers (continued)

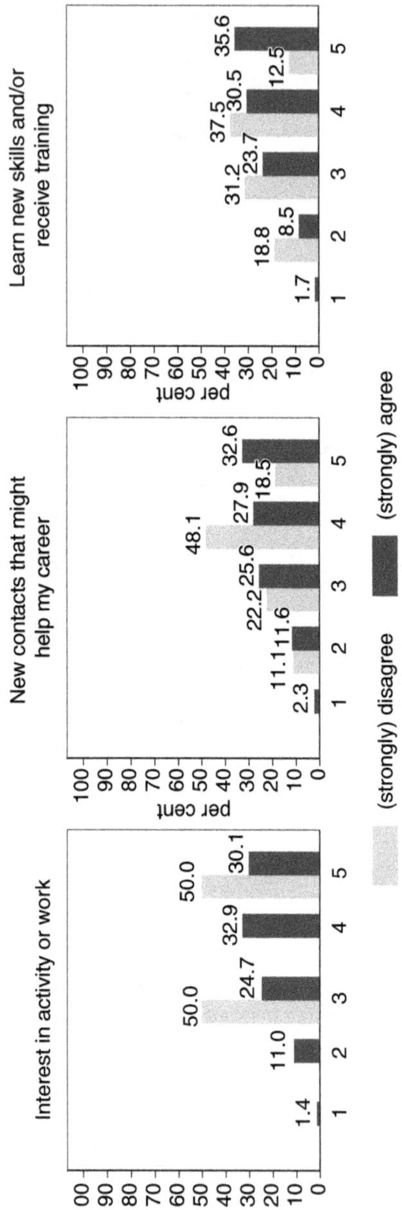

Interest in activity or work

Rating	(strongly) disagree	(strongly) agree
1	1.4	
2	11.0	
3	50.0	24.7
4	50.0	32.9
5		30.1

New contacts that might help my career

Rating	(strongly) disagree	(strongly) agree
1	2.3	
2	11.1	11.6
3	22.2	25.6
4	48.1	27.9
5	18.5	32.6

Learn new skills and/or receive training

Rating	(strongly) disagree	(strongly) agree
1	1.7	
2	18.8	8.5
3	31.2	23.7
4	37.5	30.5
5	35.6	12.5

Note: Rating of experience on a scale from one (very negative) to five (very positive)

Source: Own data and elaboration

Those CPAs who indicated agreement with egoism, in the form of emotional reward or interest in the activity, as their underlying motivation, were also rather positive in their assessment of the programme. This suggests that personal satisfaction and genuine interest in climate advocacy can be equally effective drivers of positive programme experiences. Disagreement with those two motivations, however, did not necessarily impact the personal evaluation of the programme.

For investment-driven motivation, the results are more mixed, with more people disagreeing with this factor while still rating their experience as positive. This pattern suggests that instrumental motivations, such as learning new skills or establishing new contacts, were not as consistently influential for programme satisfaction as altruistic and egoistic motives. The weaker relationship between investment motivation and positive experiences may indicate that those seeking primarily strategic benefits from their participation found the programme less aligned with their expectations than those driven by moral commitment or personal interest.

The last factor to consider as influential on the rating of the experience with the mandate is the subjective perception of one's internal as well as the external political efficacy. Figure 5.13 shows the internal and external political efficacy with regard to the experiences with the mandate. As reported earlier, due to the overall relatively high internal political efficacy values of CPAs, this does not influence their satisfaction with the mandate as much. The mean values of internal political efficacy for those who rate their experience positive or even very positive are not significantly different from those who were less satisfied.

Most CPAs estimated their external political efficacy in general as lower than their internal political efficacy. Still, based on Figure 5.13, we cannot infer that there is a significant relationship between the quality of experience with the mandate and the personal external political efficacy. For example, CPAs who rated their program experience as medium

Figure 5.13: Experience with mandate and internal as well as external political efficacy

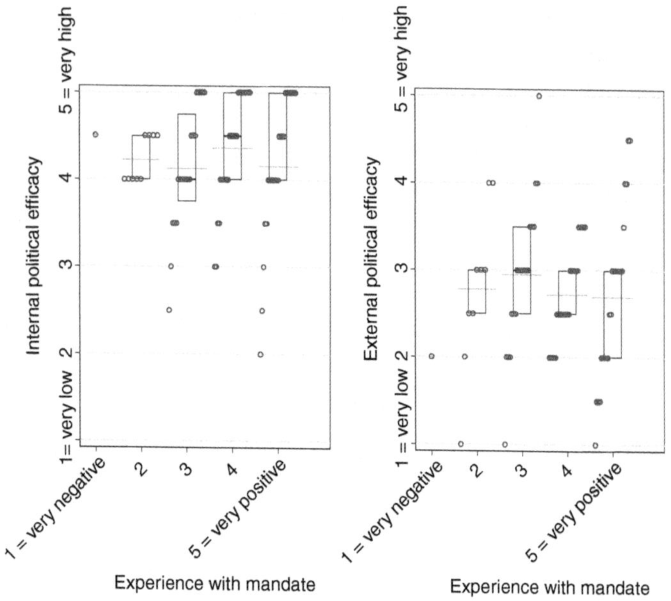

Note: Individual data points are depicted as circles, the median as a black line inside the box and the mean as a grey line

Source: Own data and elaboration

showed the highest levels of external political efficacy (both mean and upper quartile), while those reporting positive or very positive experiences demonstrated lower average political efficacy scores.

The survey also contained a field for suggesting programme changes. Of the 78 survey participants, 44 provided suggestions. The open responses cover various topics. Several CPAs indicated that pre-mandate information about benefits and limitations would be helpful. Similarly, CPAs suggested clearer guidelines regarding the European Commission's expected role. Somewhat in contrast to the impression we obtained from the responses to the closed survey questions, the responses to the open question suggested that several volunteers joined the programme to acquire new skills for more effective future activities, even after completing their mandate. Consequently, improving the training of CPAs and providing guidance on climate action implementation were other frequently mentioned issues. Another set of suggestions calls for improving communication and networking among CPAs.

One participant specifically requested encouraging CPA aggregation to shift from individual to collective actors. A concrete suggestion proposed forming thematic clusters of CPAs for greater impact. This call for better coordination is interesting because the programme emphasizes individual actors and organizational representatives. While several participants suggested member state-based coordination, thematic clusters could transcend national climate action perspectives and appeal more broadly to European citizens. However, from a pragmatic perspective, coordinating national CPAs is simpler due to shared language and understanding of what works within specific member states.

By all means, this suggestion is interesting as it has direct implications for the fundamental analytical perspective adopted in this book. One of the main points made is that the ECP's ambassador programme is innovative because it targets and empowers individuals to pursue climate action. In other words, the programme is interesting from an analytical viewpoint because it deliberately abstains from creating a collective actor that participates in climate and environmental governance. Perhaps this suggestion can be interpreted as a call to use CPAs in both capacities: as individual and collective or networked actors. The latter could be particularly useful to put forth policy demands, for example.

The mandate's duration, initially 1 year with possible annual renewals, was considered too short by some of the survey participants. This criticism warrants reflection, as individuals need time to familiarize themselves with tasks and develop networks. This is particularly relevant for building relationships with policy makers, who are busy and may take time to meet. As Tosun et al (2023b) argue, the relatively short duration of the regular term of the ambassadorship is also likely to induce CPAs to start thinking rather early in serving their mandate, whether and how they want to continue to individual-level climate action. While such a reflection can be useful for both the volunteer and the programme, it may also absorb cognitive resources and result in the CPA putting less effort into the current work for the running mandate, but planning future activities.

While the European Commission appears aware of some suggestions, certain changes may be unfeasible. Several CPAs requested European Commission affiliation or accreditation. Resource constraints were identified as a major barrier to effective climate action (see also Marqués Ruizen, 2024), alongside a lack of practical support, such as for obtaining badges for the annual UNFCCC conferences. This connects to suggestions for enhanced country coordinator roles with increased responsibility, budget, and event organization capacity. One participant proposed having an employed CPA in every EU region, supported by volunteer Friends of the Pact (see Chapter 1). However, the European Commission is unlikely to modify the programme in such a way as to shift partially from volunteering to professional CPAs, especially because this would undermine the logic on which the ECP operates. The ECP is most likely to be effective if EU citizens develop an ownership of it, and this is probably best achieved by continuing to rely on volunteering.

Several CPAs expected more opportunities for direct policy maker interactions and meaningful policy preparation involvement through formal consultations and parliamentary hearings, which provide further compelling support for the reasoning that the mandate is indeed also regarded by CPAs as a format for policy advocacy. Given that the EU Commission actively promotes policy advocacy as a programme benefit in its recruitment materials and official communications, these comments were entirely unsurprising and reflect a natural alignment between institutional promises and ambassador expectations. One particularly engaged CPA suggested including Technical Expert Group participation as a valuable addition to the programme's offerings, recognizing the technical expertise that many ambassadors bring to climate policy discussions.

Several CPAs expressed strong interest in accessing local-level policy makers beyond the traditional EU institutional framework, particularly for supporting the Covenant of Mayors for Climate and Energy[12] in implementing practical local-level climate action initiatives. This reflects the CPAs' sophisticated understanding of the EU's complex multilevel governance system and their strategic desire for expanded policy advocacy opportunities that span different levels of government. Their interest in local engagement also demonstrates recognition that climate action requires coordination across municipal, regional, national, and supranational levels to be effective.

Summary

In this chapter, we conducted three distinct analyses to examine the motivations, activities, and experiences of CPAs. First, we demonstrated that individuals become CPAs driven by various motivations, predominantly altruistic ones. However, aligned with existing volunteering literature, we could show that egoistic motivations also played an important role, while investment-driven motivations were less prominent overall.

The survey data revealed that gender differences were most pronounced in investment-related motivations, with male respondents more frequently citing both the application of existing skills and the acquisition of new ones as key motivating factors. Age differences were even less marked than gender differences, suggesting that generational factors play a relatively minor role in motivation patterns. CPAs representing organizations predominantly emphasized societal impact and social principles in their motivations, reflecting their institutional commitments. Furthermore, a professional background proved particularly important for motivational reasons to volunteer as a CPA, showing the greatest variation amongst different professional groups, specifically for investment-based motivation.

The findings demonstrated a clear predominance of intrinsic over extrinsic motivation amongst surveyed CPAs, aligning closely with core Climate Pact values of responsibility, commitment, and ambition (see Chapter 2), signalling that the ECP and its ambassador programme succeed in attracting the 'right' kind of individuals who are genuinely committed to taking action.

Second, our empirical analysis showed that CPAs effectively leverage their mandate to influence policy decisions through various forms of advocacy. They employ diverse approaches, primarily through work in partner organizations, direct engagement with politicians, and social media activism. Their advocacy efforts primarily target politicians, particularly ministers and elected MPs from governing parties, although they also engage with opposition figures, and a few sought access to courts. Our data revealed that ambassadors' professional backgrounds significantly influenced their advocacy methods and preferred channels of engagement. Furthermore, it was demonstrated that CPAs affiliated with organizations show broader engagement across different advocacy forms compared with those with individual mandates, likely reflecting greater institutional resources and networks.

Third, while most CPAs reported positive experiences with their mandate, some encountered mixed or negative experiences that provide valuable insights for programme improvement. Motivations were influential for the evaluation of the CPA experience in the sense that those who had certain clear expectations about the programme were overall more satisfied than those who did not have specific motivations driving their initial commitment. Altruistic and egoistic motives were related to higher satisfaction levels, while investment-driven motivations did not matter as much for the rating of the overall experience. While CPAs reported higher levels of internal than external political efficacy, neither measure seemed to matter for their rating of their experience with the ambassador programme. Survey respondents suggested several valuable programme improvements, including clearer organizational structures, longer mandate periods, and enhanced collaboration amongst CPAs across thematic areas and national borders to maximize their collective impact.

SIX

Conclusions and the Way Forward

Introduction

Within the EGD, the ECP represents an ambitious attempt to mobilize citizen engagement in climate and environmental action across the EU. At its heart lies an experiment in participatory governance: the CPAs. According to Marqués Ruizen (2024), who is herself a CPA, these ambassadors represent 'a very diverse bunch of people: teachers, journalists, students, politicians, lawyers, economists, city mayors or members of city councils, influencers, activists, entrepreneurs, public officials, scientists and researchers, sportspeople ...'. These individuals, drawn from diverse backgrounds across all EU member states, can be seen to embody a step towards green European citizenship (Machin and Tan, 2024).

What motivates individuals to become CPAs? Do CPAs engage in policy advocacy? If CPAs do engage in advocacy, which strategies do they apply? What experience have CPAs gained with their mandate? These questions guided this explorative study in which we analyzed original data from the European Climate Pact Ambassador Survey (Rzepka and Tosun, 2025).

The significance of understanding CPAs extends far beyond mere academic curiosity and represents a crucial area of inquiry for contemporary climate governance research. As climate change and environmental degradation increasingly demand 'accelerated' action across multiple levels of governance (Boasson et al, 2025), the role of actors who can facilitate this acceleration becomes ever more important. CPAs occupy a uniquely advantageous position within the EU's complex governance landscape, possessing both the institutional legitimacy that comes from endorsement by the European Commission and the grassroots authenticity that derives from their volunteer status and established local connections within their respective communities.

This dual positioning grants them access to both formal policy networks and informal community structures, potentially enabling them to bridge the gap between high-level climate and environmental policy objectives and local implementation realities. Given that they may play an important role in delivering on the acceleration of climate and environmental action, particularly in translating EU-level ambitions into tangible outcomes, a thorough understanding of their functions becomes essential for evaluating the broader success of European governance mechanisms.

This concluding chapter synthesizes the empirical findings of our analysis of the CPA motivations, advocacy strategies, and programme experiences. As we stand at a critical juncture in climate and environmental action, understanding how initiatives like the ECP function and how they might be enhanced becomes essential for developing effective strategies for citizen engagement in climate and environmental governance.

Climate Pact Ambassadors' motivations

In order to assess which factors motivate individuals to become a CPA, we applied Ziemek's (2006) three-category framework, which encompasses altruistic, investment, and egoistic motivational dimensions. We chose this framework because it was developed for cross-country research, which fits the nature of CPAs who are based in different EU member states, and offered a noticeably general perspective on volunteers' motivations.

The survey findings revealed that the majority of CPAs expressed strong agreement that social principles and moral obligations constitute their primary motivation for engagement in the programme. This finding underscores the ethical foundation underlying the ambassadorship role, suggesting that the survey respondents view their involvement as a moral imperative and civic duty rather than merely a voluntary activity in the narrow sense of personal interest or professional development. The strength of this ethical commitment indicates that CPAs are driven by a deeper sense of purpose that transcends individual benefits, positioning their work within a broader framework of social responsibility and collective action.

This finding proves particularly interesting and theoretically significant when considered alongside the analysis of CPAs' mission statements published on the ECP online platform by Tosun et al (2023a). While this research anticipated that a sense of responsibility, particularly by young people towards younger and future generations, would serve as a key motivator for individuals to volunteer as climate ambassadors, we were unable to demonstrate this empirically based on the available data sources. The survey results provide empirical support for the importance of this sense of responsibility and moral duty. Notably, even though a marked age effect among respondents could not be determined, which aligns with the findings of Tosun et al (2023a), the moral imperative appears to transcend demographic boundaries and unite CPAs across different age cohorts in their shared commitment to climate action.

The second most prominent motivational factor was the recognition of societal need, specifically, the belief that there is a demand for individuals willing to undertake this mandate. This response pattern suggests that CPAs are motivated not only by personal convictions but also by their assessment of gaps in climate action that require citizen engagement. This response reinforces the altruistic foundation of CPA engagement. It suggests that volunteers are motivated by a sense of duty, which aligns with the reflections by Machin and Tan (2024) on the relationship between the EGD and green European citizenship.

Beyond these primary drivers, this analysis revealed several secondary motivational factors that contribute to CPA engagement. Numerous respondents cited general interest in climate-related work as a motivating factor, indicating that intrinsic motivation to engage with environmental issues serves as an important catalyst for involvement. This motivation suggests that participation formats, such as the ECP, have the potential to mobilize people to become engaged in actions addressing climate change and the environment.

The emotional rewards derived from acting as a CPA emerged as another important motivational factor that underpins sustained engagement with the programme. Survey respondents frequently highlighted the personal fulfilment and sense of accomplishment they experienced through their ambassadorial activities. The emotional satisfaction of this volunteering contribution could create a positive feedback loop that sustains long-term volunteer commitment, which, in turn, reinforces their dedication to the role and encourages continued participation in programme activities (see Omoto and Snyder, 2002).

Noteworthy is also the number of CPAs who indicated that skill development and training opportunities motivate their participation. This finding suggests that the ambassadorship serves a dual purpose: advancing climate and environmental action while simultaneously providing professional development opportunities for volunteers. The integration of learning and service creates value for both the individual volunteer and the broader climate and environmental movement, potentially enhancing both personal capacity and programme effectiveness. From this perspective, the ambassadorship creates a win-win situation for stakeholders in climate and environmental governance.

Conversely, relatively few CPAs indicated that networking or career advancement opportunities motivated their participation. This finding is interesting because it suggests that the ambassadorship attracts individuals driven primarily by substantive rather than instrumental concerns. The low importance placed on career networking indicates that the CPA programme attracts volunteers with a genuine commitment to climate and environmental issues rather than those seeking professional advancement. However, it must be noted that the recruitment criteria require volunteers to be leaders in their communities and thus to be already well-connected to other people. Furthermore, the European Commission selects who to appoint as a CPA, and aspirants conveying the impression that they primarily wish to pursue their career through the programme are likely to be filtered out during this selection process.

The aforementioned findings referred to the prevalence of the answers given by the respondents. In a second approach, the CPAs were asked to identify their single most important motivational factor, which revealed an interesting hierarchy of motivations. The paramount motivation indicated by the CPAs was society's need for committed individuals willing to foster climate action, which aligns with the responsibility-focused understanding of green European citizenship by Machin and Tan (2024).

The second most important factor was personal interest in climate issues, followed by social and moral principles, and the desire to utilize existing skills and experience. Other factors, including community connections, emotional rewards, and feelings of obligation, were cited as primary motivators by relatively few individuals, although they remain important secondary considerations. The importance of personal interest in the issue may be the outcome of the high levels of attention for climate change at the time when the survey was fielded and when the CPAs decided to volunteer for the programme.

Several factors explain these motivational differences among CPAs, with professional background being the primary influence. CPAs from different sectors emphasized distinct reasons for engagement – for instance, students highlighted their desire to build new professional networks. At the same time, those with an NGO background brought a different perspective to their motivation. Having witnessed firsthand the challenges facing organizations working on climate and environmental issues, they mostly felt that there was a great need for this work. The most important motivational dimension of altruism was relevant to all CPAs, no matter their age, gender, or type of mandate. Gender differences were especially prominent for the investment-driven motivation to use skills, which was mostly stated by men between their thirties and their sixties. Those representing organizations with their mandate mainly emphasized social principles in their motivations.

Overall, the findings on the CPAs' motivation align with Ziemek's (2006) three-category framework and demonstrate that established concepts could be used for assessing the motivations underlying the very specific type of volunteering represented by the ambassador programme, despite its focus on climate and environmental issues and the multilevel character of the format that involves submitting an application to the European Commission in order to receive a mandate for local and community level engagement.

This multidimensional motivational profile suggests that effective ambassador programmes should appeal to diverse motivational drivers rather than focusing solely on altruistic appeals and ensure that they also offer ample opportunities for learning and professional development. The complexity of motivational factors indicates that CPA programmes serve multiple functions within the broader climate movement: they provide channels for civic engagement, create learning and development opportunities, and offer pathways for individuals to align their interests with societal needs. Programme developers are well advised to take these multiple functions into consideration.

Climate Pact Ambassadors' engagement in policy advocacy

As discussed in Chapter 2, the European Commission assigns CPAs several roles, which have in common that they involve communication activities. This feature of the ambassadorship also aligns with the etymological meaning of the term as highlighted by Reas et al (2023). This broad understanding of the roles of CPAs opens up opportunities for both climate and environmental activism and advocacy (Parks et al, 2023). This reasoning is supported by Machin and Tan (2024), who, based on the Commission's Communication (2020), allude to the ECP calling on individuals to participate in discussions, also with the explicit goal of improving decision-making processes. In this interpretation, this includes policy advocacy, which also aligns with the information provided on the European Commission's website on the ECP (European Commission, 2025b). The European Commission's emphasis on communication as a core component of the CPA role suggests that these positions are designed not merely as ceremonial appointments but as active conduits for public engagement and policy influence within the European climate governance framework.

The survey data revealed that CPAs indeed endeavour to shape climate and environmental policy. Most CPAs worked in another organization to broaden their policy advocacy in further forums. Many also posted about politics online, signed a petition, or contacted politicians. This indicates that policy advocacy constitutes one of several activities CPAs carry out – precisely as envisaged by the European Commission when establishing these roles.

The advocacy strategies differ markedly between individual CPAs and those representing organizations, with organization-mandated CPAs demonstrating overall higher rates of participation in policy advocacy activities. The most common form of policy advocacy is to work within another organization, reported by both organization-mandated and individual CPAs alike, such as an NGO working on climate and environmental issues.

Less common advocacy activities among both individual and organizational CPAs include working actively in political parties, wearing campaign badges, and participating in demonstrations or protests. The latter is particularly interesting in light of the argument put forth by Parks et al (2023) that we need to distinguish between activism and advocacy as distinct forms of political engagement. In light of these findings, CPAs can be seen as preferring advocacy approaches over activism. At the same time, some of them also engage in activism, that is, in participating in protests and demonstrations that involve more direct action.

The type of policy advocacy pursued varies considerably according to the CPAs' professional background and sectoral experience. Academic CPAs primarily focus on political outreach and online political engagement across an array of digital platforms and social media channels. CPAs based in the education sector tend to favour working in a political party and posting about politics online. All CPAs from civil society and NGOs primarily work through other organizations, leveraging existing networks and collaborative relationships. Besides, compared with the other occupational groups, they had the highest share of contacting politicians. CPAs based in the private sector show high engagement across multiple activities, including seeking access to politicians or signing petitions. Students predominantly share political statements online, possibly reflecting their greater familiarity with social media and online presence in general, but also indicating the more limited advocacy strategies at their disposal. These patterns clearly demonstrate that a professional background influences the types of policy advocacy CPAs pursue during their mandate period.

When analyzing the targets of CPA advocacy strategies and looking at their frequencies of contacting distinct political stakeholders, we showed that a higher share of organization-mandated CPAs contacted different target groups compared with their individual-mandated counterparts, demonstrating greater flexibility in their approach. CPAs primarily sought access to ministers and MPs from governing or majority parties, with considerably less focus on opposition or minority party MPs and courts. This pattern indicates that CPAs prioritize engaging with influential politicians who can directly affect agenda-setting and policy-making processes, and it supports the view that these kinds of activities are indeed undertaken as genuine efforts for advocating certain policies rather than merely fulfilling formal obligations as part of the ambassadorship.

The access to various political stakeholders is not only influenced by their mandate, but also the CPAs' professional background. Most notably, most CPAs working in NGOs accessed ministers and elected MPs from governing parties, while those affiliated with the public sector focused, aside from contacting predominantly ministers, on courts. Hence, CPAs demonstrated developing a targeted approach to pursue their policy advocacy depending on their resources, networks, knowledge, and mandate stemming from their working environment.

While the literature on volunteering recognizes volunteers as policy advocates (Snyder and Omoto, 2008; Nesbit, 2017), discussing these findings in light of this literature is challenging since it does not utilize the types of advocacy activities our research draws from, which originates from Beyers et al (2020). However, because of the conceptualization of advocacy activities, these findings relate to research on interest groups and policy advocacy. From that viewpoint, the conceptual approach provided by that literature lends itself well to studying the activities of CPAs. This alignment suggests that CPAs function similarly to other policy actors in the European political system, albeit with the legitimacy and access that comes from their official Commission-endorsed status.

Climate Pact Ambassadors' experience with the programme

More than half of the CPAs who participated in the survey indicated that their experience with the programme was (very) positive, followed by approximately 26 per cent reporting a mixed experience, and around 13 per cent describing their experience as (very) negative. Assuming the responses were provided sincerely, this overall positive assessment suggests that the programme has been largely successful in meeting participants' expectations and providing meaningful engagement opportunities, although room remains for improvement across various aspects of the initiative.

To explain the variation in CPAs' experiences with the programme, we focused on the motivation of CPAs to pursue this mandate, as well as on their perceived internal and external political efficacy. Those who started to volunteer in this programme with a clear motivation were also more often satisfied with the programme than those who disagreed with certain motivational factors. Nevertheless, motivation sometimes showed mixed results, especially for investment-driven motivational statements.

Finally, how well CPAs perceived their political efficacy did not matter for their rating as much as expected. Because a high share of CPAs reported high levels of internal political self-efficacy, there was not enough variation to deduce effects on the rating of the experience from this. CPAs with a medium level of external political efficacy were on average more satisfied than those who reported higher levels of external political efficacy, showing mixed results for this item. This may be explained by the fact that the tangible impact of the ambassador programme on climate and environmental policy and governance is more limited than initially expected when volunteering for the role.

The survey also yielded valuable insights into how the programme could be improved from the participants' perspective, offering concrete suggestions, some of which align with those proposed by Marqués Ruizen (2024). Several CPAs suggested that, prior to applying for the role, it would be helpful to receive comprehensive information about the benefits of the mandate, as well as a realistic assessment of the programme's limitations and potential challenges. This would help manage expectations and ensure that volunteers have a clear understanding of what the role entails and what can realistically be achieved.

In this context, CPAs called for clearer guidance regarding the role the European Commission expects them to play, indicating some uncertainty about their responsibilities, scope of action, and the boundaries of their mandate. Such definitional challenges suggest that role clarity could be relevant for enhancing both ambassador satisfaction and programme outcomes.

Another frequently mentioned issue concerned the training of CPAs and the provision of practical guidance on implementing climate action effectively within their respective contexts. Comments indicated that many volunteers viewed the programme as an opportunity to learn new skills and build their capabilities. This aspiration also emerged in the earlier section on CPAs' motivations for seeking the mandate. Regarding this point, it is worth noting that the ECP Secretariat has responded by expanding training opportunities, including the creation of new institutions such as the EU Climate Action Academy.

A further set of suggestions centred on improving communication and networking among CPAs, highlighting the importance of fostering stronger peer connections, creating more opportunities for collaboration, and facilitating the sharing of knowledge and best practices between ambassadors from diverse regions and backgrounds. These recommendations reflect a broader desire for enhanced interaction and mutual learning within the ambassadorial community. With this in mind, and taking into account that the data were collected in late 2023, the ECP Secretariat has since undertaken efforts to improve communication among CPAs through a range of new initiatives and digital platforms.

Looking ahead, when collecting data from CPAs who joined the programme after 2023, it will be particularly insightful to evaluate whether concerns about communication and connectivity persist or whether these recent improvements have been effective in addressing the issues previously raised, thereby strengthening the collaborative and community-building aspects of the programme.

Practical implications of this research

The principal contribution of this study to policy practice centres on the design and operation of the ambassador programme itself. The findings reveal areas where the programme's current structure may be falling short of its intended objectives, while simultaneously highlighting pathways for enhancement that could strengthen both the programme's effectiveness and its contribution to the EU's legitimacy.

The evidence presented in this research demonstrates that approximately half of the ambassadors participating in the programme express some level of dissatisfaction with their experience. This level of discontent represents a concern for the EU Commission, particularly given the programme's role in fostering citizen engagement and democratic participation within the EU. However, rather than viewing this dissatisfaction as an insurmountable obstacle, it should be understood as valuable feedback that illuminates specific areas requiring attention and reform.

Importantly, many of the ambassadors who expressed dissatisfaction do not merely voice complaints but contribute constructive suggestions for improvement (see Chapter 5). These recommendations, emerging from direct participant experience, provide invaluable insights into the practical challenges faced by those engaging with the programme on the ground. This feedback represents a rich source of knowledge that can inform evidence-based programme adjustments, ensuring that the ECP is more responsive to participant needs and expectations in the future, as the increase in the number of CPAs (see Chapter 2) clearly demonstrates that there is sustained interest in this format.

The significance of responding to these concerns extends beyond the immediate scope of programme improvement. By demonstrating a genuine commitment to learning from participant feedback and implementing meaningful changes, the EU Commission and the ECP Secretariat can showcase their willingness to adapt and evolve their participatory governance mechanisms. This responsiveness to citizen input serves as a powerful signal that the EU values democratic engagement and takes seriously its responsibility to create effective channels for citizen participation in European governance. In fact, the ECP Secretariat has demonstrated its capacity and willingness to adapt the programme based on both its own experience and feedback from CPAs (see Chapter 2).

Moreover, strengthening the ambassador programme based on participant feedback can contribute to broader perceptions of EU legitimacy. When citizens observe that their voices are heard and that their suggestions lead to improvements, it reinforces the notion that the European Commission is a responsive and accountable institution and that it delivers on its commitment to co-production (European Commission, 2018).

Ultimately, the enhanced programme design can be expected to lead to more positive participant experiences, which in turn can be expected to strengthen citizen trust in EU institutions and their commitment to democratic participation. This increased legitimacy provides a stronger foundation for further participatory initiatives, creating a self-reinforcing dynamic that can contribute to the long-term sustainability of the EU's democratic governance.

Future research directions

This book offers the first systematic discussion of the ECP in general and of the CPAs' motivations, strategies, and experiences based on original empirical research in particular. It also represents an attempt to integrate different literatures that have developed in isolation. Despite the merits of this research, the analysis presented in this book could be developed further in several ways, and future research is encouraged to consider the following potential avenues.

Conceptually, in this book, we concentrated on volunteering as well as climate and environmental activism and advocacy. Of these two latter areas, we paid more attention to advocacy and embraced an operationalization of activism that is compatible with advocacy as provided by Beyers et al (2020). However, more types of activism exist, and most importantly, more contentious ones than those captured here (Fisher and Nasrin, 2021; Foxe et al, 2024). Future research might therefore benefit from exploring how CPAs relate to these more confrontational forms of climate activism and whether their institutional positioning influences their willingness to engage with or support more radical climate action strategies.

In addition to a more comprehensive assessment of activism, we are aware that we could have applied alternative conceptual lenses to assess the activities in which CPAs engage. For example, we did not assess how CPAs participate in governance processes, although the EU context offers many such formats, as part of the EGD (see Chapter 2) and beyond. Along similar lines, it was beyond the scope of this study to show how CPAs participate in multilevel governance, that is, their participation in governance arrangements at different levels of the EU's political system (Newig and Koontz, 2014).

Focusing on the concept of multilevel governance more narrowly, it would be interesting for future research to examine at which level of the political system CPAs are active and (more) successful in mobilizing citizens to participate in climate and environmental action. Based on the data we consulted for this research, the impression is that CPAs are mostly active at the community level, which resonates with the programme's design and intended scope. Still, this impression warrants a more systematic assessment, especially because we have also seen that the targets of the CPAs' advocacy efforts are also decision makers at the national and the EU level.

Likewise, in order to gain a more comprehensive understanding of how CPAs engage in governance processes, social network analysis would constitute a particularly suitable and insightful methodological approach (Berardo et al, 2020; Wagner et al, 2023). This technique allows for the examination of the structure and dynamics of relationships among actors, shedding light on how information, resources, and influence flow within and across networks. Consequently, a productive avenue for future research building on the present analysis would be to investigate the social networks of CPAs in a more systematic and detailed manner. This could involve mapping their connections, identifying key nodes of influence, and analyzing the nature and strength of their ties across various levels of governance: local, national, and supranational.

From a theoretical viewpoint, we did not pay attention to whether CPAs had already collected experience as volunteers before applying for the programme. It is possible that individuals who have already volunteered or who have participated in other formats for participatory or deliberative governance, such as climate juries or climate citizen assemblies (Lorenzoni et al, 2025), are more likely to volunteer as CPAs. This prior experience might influence their effectiveness and approach to their ambassadorial role.

Conversely, it is also interesting to learn what CPAs do after completing their mandate, a perspective suggested by Tosun et al (2023b). More generally, value could be seen in understanding the participation patterns of CPAs with regard to their previous engagement and subsequent activities after completing their term. Thus, a longitudinal analysis would be helpful to understand whether the ambassador's programme is the cause of future engagement and therefore can have a more sustained impact on individuals' civic participation trajectories.

As already noted in Chapter 3, another major limitation of this analysis is that we did not test a theoretical model. Instead, we provided insights for certain elements of theoretical models. Given the limitations imposed by the data, the analysis is mostly of an explorative nature, and we consider the findings reported as indicative and preliminary. Future researchers are invited to critically reflect on the model proposed in this book, refine or revise it if they consider it necessary, and test hypotheses derived from it using more robust methodological approaches.

This takes us to the empirical foundation of this analysis, which would have benefitted from strengthening the data basis. Chapter 4 elaborated on how the survey respondents were recruited. The recruitment relied predominantly on using email invitations, which resulted in a relatively small sample size that limited the generalizability of the findings. Future research may use other ways of recruiting respondents to achieve greater representativeness. For example, the annual meeting of CPAs could be an excellent event to present future research projects focusing on the ECP and CPAs and to invite attendees to participate in surveys or workshops. Moreover, it could be helpful for the recruitment to offer the survey respondents incentives.

In addition to the sampling of respondents, the theoretical constructs underlying the empirical analysis can be modified, expanded, and refined. As mentioned previously, examining CPAs from a multilevel governance perspective would require different survey questions to better capture the corresponding governance dynamics. Along similar lines, and as argued in Chapter 3, the survey questionnaire could include questions to capture all elements of the three-stage Volunteer Process Model by Omoto and Snyder (2002), providing a more comprehensive theoretical foundation and a deeper understanding of volunteer retention determinants.

Furthermore, an alternative analytical lens we could have applied to CPAs is that of climate intermediaries (see Chapter 2). While a rich literature exists on intermediation (Abbott et al, 2017), also on their role for sustainability transitions (Kivimaa et al, 2020), only recently has this concept been applied to climate policy research defined narrowly (Tobin et al, 2023b; Tosun et al, 2023c). From this perspective, a promising angle for future research can be seen in more systematically conceiving of CPAs as climate intermediaries and assessing how they perform their role as 'go-betweens' and how citizens and governmental actors assess their performance in this capacity.

Likewise, it would be instructive to assess whether CPAs perceive themselves as intermediaries and how this self-perception influences their approach to their ambassadorial duties. This line of inquiry could provide valuable insights into the mechanisms through which CPAs facilitate communication and action between different levels of governance and civil society.

Taken together, the findings presented in this research open up multiple promising avenues for future enquiry. Some of these avenues could focus more specifically on the role and impact of CPAs within the broader framework of climate governance. In contrast, others might explore different forms of climate and environmental activism and advocacy, including how various actors engage with policy processes, mobilize communities, or influence public discourse on sustainability issues. Still other avenues could involve applying different analytical frameworks to CPAs, such as examining them through the lens of multilevel governance or as climate intermediaries, as suggested above.

Future outlook

This research centred on the EGD as it was launched in 2019. In numerous respects, the reality within the EU proved to be markedly different from initial expectations and policy projections. Since the inception of the EGD, the EU, much like other regions, has been compelled to wrestle with the unprecedented COVID-19 pandemic and its far-reaching social and economic ramifications. It also had to deal with the consequences of Russia's attack on Ukraine in February 2022. The latter has fundamentally altered several social, economic, and political parameters across the EU, creating a new geopolitical reality that policy makers had not anticipated (von Homeyer et al, 2022; Tosun, 2023). Most importantly, these events have contributed to a prolonged period of economic stagnation in the EU, creating pressure on EU and national politicians to re-emphasize the economic growth orientation of the EGD.

In response to this, the European Commission published the Clean Industrial Deal (CID) in February 2025, which endeavours to combine the decarbonization of the European industry with the strengthening of its competitiveness and innovative capacity, while simultaneously improving economic resilience and the security of supply chains (European Commission, 2025d). This new framework represents a shift towards a more pragmatic approach to climate policy. While the CID remains committed to achieving climate-neutrality by the previously established targets, it offers a more strategically focused and economically viable approach that aims to restore the EU's international competitiveness in an increasingly complex and volatile global marketplace. It remains to be seen what precise role the ECP will play in this second phase of the EGD's evolution and implementation.

The fact that at the time of writing, the ambassador programme has attracted more than 1,100 individuals, representing the highest number since its initial launch, constitutes an encouraging observation for continued citizen engagement and grassroots support. This suggests that despite the challenging circumstances, a core group of committed individuals remains, who continue to prioritize climate and environmental action. Nonetheless, considering the overall challenging situation facing the EU, including economic pressures and security concerns, it will be particularly interesting to observe whether the current level of citizen engagement targeting climate action and environmental protection will remain stable over time, or whether external pressures and competing priorities will influence public sentiment, participation rates, and long-term commitment to the goals of the EGD.

Notes

one Introducing the European Climate Pact and its Ambassadors

[1] It should be noted that while the EU maintained its overall policy effort, research by Burns and Tobin (2020) and Brandsma et al (2023), for instance, has also alluded to instances of waning levels of policy ambition during times of crises and instances of policy dismantling and termination.

two Climate Action Through the Lens of the European Climate Pact

[1] According to https://climateambassadors.org.uk/, in the UK, a Climate Ambassadors scheme exists that aims to provide expertise to educational institutions (nurseries, schools, and colleges) in order to develop and implement effective climate action plans. This scheme is led by the University of Reading and the Alliance for Sustainability Leadership in Education and is funded by the Department for Education. The programme operates nine regional hubs that support and connect the climate ambassadors and educational settings. It is delivered by a consortium of 15 organizations, including eight universities.

[2] This information was obtained from: https://climatechampions.unfccc.int/global-ambassadors/

[3] This information was obtained from: https://climate-pact.europa.eu/eu-climate-action-academy/webinars-and-training_en

[4] This information was obtained from: https://ec.europa.eu/commission/presscorner/detail/en/ip_17_1872

[5] The website can be accessed via this link: https://citizens-initiative.europa.eu/_en

[6] The Climate Action Tracker is an independent scientific project to assess the progress countries make in meeting their commitments under the Paris Agreement. The website can be accessed via this link: https://climateactiontracker.org/

NOTES

four Research Design and Operationalization

1. Focusing on grocery store customers who are asked to adopt an ambassador role regarding the use of reusable shopping bags, Hassler et al (2025) show that, alongside private aspects, the perception of the publicness of this ambassador role varies across individuals and that those who are more aware of this aspect are more likely to use renewable bags.
2. The dedicated website with the public profiles can be found at: https://climate-pact.europa.eu/meet-community/climate-pact-ambassadors_en
3. The design of the online survey and its administration was a joint work with David Rzepka as part of his Bachelor's thesis prepared at the Institute of Political Science at Heidelberg University. The survey adhered to Heidelberg University's data policy, including its stipulations for data protection. The respondents provided their informed consent for the use of the data for research purposes and publishing findings based on this research.
4. This figure corresponds to the information provided on 4 July 2025, available at: https://climate-pact.europa.eu/meet-community/climate-pact-ambassadors_en
5. The dataset and codebook can be accessed at: https://doi.org/10.11588/DATA/ZIDN7D

five Empirical Findings

1. Although the ECP website displays all CPAs with their names, the names were not stated directly here to avoid drawing undue attention to specific CPAs and putting them on the spot.
2. This information was obtained from: https://ec.europa.eu/newsroom/clima/newsletter-archives/56146?utm_source=chatgpt.com
3. This information was obtained from: https://ec.europa.eu/newsroom/clima/newsletter-archives/56146?utm_source=chatgpt.com
4. This information was obtained from: https://jpi-climate.eu/news/blog-climate-pact-ambassador-for-the-eu-jpi-climate-members-involved/
5. This information was obtained from: https://inzeb.org/projects/european-climate-pact/
6. This information was obtained from: https://ec.europa.eu/newsroom/clima/items/871285/en
7. Peer parliaments are a tool established by the ECP. It involves bringing together a small group of people and discussing with them how the transition to climate neutrality can work in practice and what policies should be put in place to encourage us going forward. The target groups are people with whom the CPAs have close relations or a targeted

group in their community. The CPAs can use a toolkit for facilitating peer parliaments, which are available on the ECP website. For further information, consult the following website: https://climate-pact.europa.eu/get-involved/host-group-activity/quick-start-tools-citizen-engagement/peer-parliament_en

[8] This information was obtained from:
https://ec.europa.eu/newsroom/clima/items/866753/en

[9] This information was obtained from: https://www.linkedin.com/posts/klimatkommunerna_euclimatepact-myworldourplanet-activity-7336293226351419393-gUxY/?utm_source=share&utm_medium=member_desktop&rcm=ACoAACIBo54BncmPwCPsEc2bHqCwMYNbzunHPWM

[10] This information was obtained from: https://climate-pact.europa.eu/meet-community/climate-pact-ambassadors/marta-fandlova_en

[11] This information was obtained from: https://climate-pact.europa.eu/articles-and-events/events/european-climate-pact-flagship-event-2025-together-action-2025-03-19_en

[12] Founded in 2008, the Covenant of Mayors for Climate and Energy is the mainstream European movement involving local authorities in the development and implementation of sustainable energy and climate policies. In 2017, the Covenant of Mayors for Climate and Energy and Compact of Mayors merged into the Global Covenant of Mayors for Climate and Energy, an international alliance of local governments (Francesco et al, 2020). Since then, the Covenant of Mayors for Climate and Energy continues to represent the European branch of the international network.

References

Abbott, K.W., Levi-Faur, D. and Snidal, D. (2017) 'Theorizing regulatory intermediaries', *The Annals of the American Academy of Political and Social Science*, 670(1): 14–35.

Ackermann, K. (2019) 'Predisposed to volunteer? Personality traits and different forms of volunteering', *Nonprofit and Voluntary Sector Quarterly*, 48(6): 1119–42.

Alemanno, A. (2021) *Strengthening the Role and Impact of Petitions as an Instrument of Participatory Democracy: Lessons Learnt from a Citizens' Perspective 10 Years after the Entry into Force of the Lisbon Treaty*. European Parliament.

Almeida, P., González Márquez, L.R. and Fonsah, E. (2024) 'The forms of climate action', *Sociology Compass*, 18(2): e13177.

Arabadjieva, K. and Bogojević, S. (2024) 'The European Green Deal: climate action, social impacts and just transition safeguards', *Yearbook of European Law*, 43(1): 34–55.

Australian Bureau of Statistics (2021) '*General Social Survey 2020 Questionnaire*', [online] Available from: https://www.abs.gov.au/methodologies/general-social-survey-summary-results-australia-methodology/2020 [Accessed 14 August 2025].

Bale, T., Webb, P. and Poletti, M. (2019) *Footsoldiers: Political Party Membership in the 21st Century*. Routledge.

Bandura, A. (1997) *Self-Efficacy: The Exercise of Control*. Freeman.

Beierlein, C., Kemper, C.J., Kovaleva, A. and Rammstedt, B. (2012) 'Ein Messinstrument zur Erfassung politischer Kompetenz-und Einflussüberzeugungen: Political Efficacy Kurzskala (PEKS)', *GESIS-Leibniz-Institut für Sozialwissenschaften Working Papers*, (18): 1–24.

Berardo, R., Fischer, M. and Hamilton, M. (2020) 'Collaborative governance and the challenges of network-based research', *The American Review of Public Administration*, 50(8): 898–913.

Beyers, J., Fink-Hafner, D., Maloney, W.A., Novak, M. and Heylen, F. (2020) 'The Comparative Interest Group-survey project: Design, practical lessons, and data sets', *Interest Groups & Advocacy*, 9(3): 272–89.

Biesbroek, R., Peters, B.G. and Tosun, J. (2018) 'Public bureaucracy and climate change adaptation', *Review of Policy Research*, 35(6): 776–91.

Binderkrantz, A. (2008) 'Different groups, different strategies: How interest groups pursue their political ambitions', *Scandinavian Political Studies*, 31(2): 173–200.

Binderkrantz, A.S., Blom-Hansen, J., Baekgaard, M. and Serritzlew, S. (2023) 'Stakeholder consultations in the EU Commission: instruments of involvement or legitimacy?', *Journal of European Public Policy*, 30(6): 1142–62.

Boasson, E.L., Peters, G.P. and Tosun, J. (2025) 'Policy-driven acceleration of climate action', *PLOS Climate*, 4(5): e0000626.

Boasson, E.L. and Tatham, M. (2023) 'Climate policy: From complexity to consensus?' *Journal of European Public Policy*, 30(3): 401–24.

Bocquillon, P. (2024) 'Setting the European agenda in hard times: The commission, the European Council and the EU polycrisis', *Journal of European Integration*, 46(4): 567–74.

Borbáth, E. and Hutter, S. (2024) 'Environmental protests in Europe', *Journal of European Public Policy* (early view): 1–26.

Bouza Garcia, L. (2010) 'From civil dialogue to participatory democracy: The role of civil society organisations in shaping the agenda in the debates on the European Constitution', *Journal of Contemporary European Research*, 6(1): 85–106.

Brandsma, G.J., Pollex, J. and Tobin, P. (2023) 'Overlooked yet ongoing: Policy termination in the European Union', *Journal of Common Market Studies*, 61(5): 1360–76.

Burns, C. and Tobin, P. (2020) 'Crisis, climate change and comitology: Policy dismantling via the backdoor?' *Journal of Common Market Studies*, 58(3): 527–44.

Buzogány, A., Parks, L. and Torney, D. (2025) 'Democracy and the European Green Deal', *Journal of European Integration*, 47(2): 135–54.

Çelik, F.B. (2025) 'Unpacking democratic participation in the European Green Deal: The case of Climate Pact', *Journal of European Integration*, 47(2): 173–92.

REFERENCES

Challies, E., Newig, J., Kochskämper, E. and Jager, N.W. (2017) 'Governance change and governance learning in Europe: Stakeholder participation in environmental policy implementation', *Policy and Society*, 36(2): 288–303.

Chilvers, J., Pallett, H. and Hargreaves, T. (2018) 'Ecologies of participation in socio-technical change: The case of energy system transitions', *Energy Research & Social Science*, 42: 199–210.

Cifuentes-Faura, J. (2022) 'European Union policies and their role in combating climate change over the years', *Air Quality, Atmosphere, & Health*, 15(8): 1333–40.

Clary, E.G. and Snyder, M. (1998) 'Understanding and assessing the motivations of volunteers: A functional approach', *Journal of Personality and Social Psychology*, 74(6): 1516–30.

Clary, E.G. and Snyder, M. (1999) 'The motivations to volunteer', *Current Directions in Psychological Science*, 8(5): 156–9.

Cnaan, R.A. and Amrofell, L. (1994) 'Mapping volunteer activity', *Nonprofit and Voluntary Sector Quarterly*, 23(4): 335–51.

Cnaan, R.A., Handy, F. and Wadsworth, M. (1996) 'Defining who is a volunteer: Conceptual and empirical considerations', *Nonprofit and Voluntary Sector Quarterly*, 25(3): 364–83.

Cnaan, R.A., Meijs, L., Brudney, J.L., Hersberger-Langloh, S., Okada, A. and Abu-Rumman, S. (2022) 'You thought that this would be easy? Seeking an understanding of episodic volunteering', *Voluntas*, 33(3): 415–27.

Cornwall, A. (2000) *Beneficiary, Consumer, Citizen: Perspectives on Participation for Poverty Reduction*, Swedish International Development Cooperation Agency.

Crespy, A. and Parks, L. (2019) 'The European Parliament and civil society', in O. Costa (ed) *The European Parliament in Times of EU Crisis*: pp 203–23.

Crum, B. (2024) 'Models of EU constitutional reform: What do we learn from the Conference on the Future of Europe?' *Global Constitutionalism*, 13(2): 392–410.

Della Porta, D. and Felicetti, A. (2022) 'Innovating democracy against democratic stress in Europe: Social movements and democratic experiments', *Representation*, 58(1): 67–84.

Dolnicar, S. and Randle, M. (2007) 'What motivates which volunteers? Psychographic heterogeneity among volunteers in Australia', *Voluntas,* 18(2): 135–55.

Driscoll, D. (2023) 'Populism and carbon tax justice: The wellow vest movement in France', *Social Problems*, 70(1): 143–63.

Dunn, J., Chambers, S.K. and Hyde, M.K. (2016) 'Systematic review of motives for episodic volunteering', *Voluntas*, 27(1): 425–64.

Dupont, C., Moore, B., Boasson, E.L., Gravey, V., Jordan, A., Kivimaa, P., Kulovesi, K. et al (2024) 'Three decades of EU climate policy: Racing toward climate neutrality?', *WIREs Climate Change*, 15(1): 1–12.

Dür, A. and Mateo, G. (2013) 'Gaining access or going public? Interest group strategies in five European countries', *European Journal of Political Research*, 52(5): 660–86.

European Commission (2001) *European Governance A White Paper.* European Commission.

European Commission (2018) *Co-production – Enhancing the Role of Citizens in Governance and Service Delivery: Technical Dossier no. 4.* European Commission.

European Commission (2019) *Communication from the Commission to the European Parliament, the European Council, the Council, the European Economic and Social Committee and the Committee of the Regions: The European Green Deal COM/2019/640 final.* European Commission.

European Commission (2020) *Communication from the Commission to the European Parliament, the Council, the European Economic and Social Committee and the Committee of the Region.* European Commission.

European Commission (2025a) 'About the Pact', [online] Available from: https://climate-pact.europa.eu/about/about-pact_en [Accessed 1 July 2025].

European Commission (2025b) Become a Pact Ambassador, [online] Available from: https://climate-pact.europa.eu/get-involved/become-pact-ambassador_en [Accessed 28 June 2025].

European Commission (2025c) Climate Pact Ambassadors, Available from: https://climate-pact.europa.eu/meet-community/climate-pact-ambassadors_en [Accessed 8 July 2025].

REFERENCES

European Commission (2025d) *The Clean Industrial Deal: A Joint Roadmap for Competitiveness and Decarbonisation: COM/2025/85 final*, European Commission.

European Commission (2025e) *European Citizens' Panels: A New Phase of Citizen Engagement,* Available from: https://citizens.ec.europa.eu/european-citizens-panels_en [Accessed 8 July 2025].

European Commission (2025f) *Futurium: Your Voices, Your Future*, Available from: https://futurium.ec.europa.eu/en [Accessed 8 July 2025].

European Social Survey ERIC (2017) European Social Survey (ESS), Round 8 – 2016.

Fisher, D.R. and Nasrin, S. (2021) 'Climate activism and its effects', *WIREs Climate Change*, 12(1): e683.

Foxe, J., Dolšak, N. and Prakash, A. (2024) 'Varieties of climate activism: assessing public support for mainstream and unorthodox climate action in the United Kingdom', *Environmental Research Communications*, 6(11): 111006.

Francesco, F.D., Leopold, L. and Tosun, J. (2020) 'Distinguishing policy surveillance from policy tracking: transnational municipal networks in climate and energy governance', *Journal of Environmental Policy & Planning*, 22(6): 857–69.

Fransen, T., Meckling, J., Stünzi, A., Schmidt, T.S., Egli, F., Schmid, N. et al (2023) 'Taking stock of the implementation gap in climate policy', *Nature Climate Change*, 13(8): 752–5.

Gaventa, J. (2006) 'Finding the spaces for change: A power analysis', *IDS Bulletin*, 37(6): 23–33.

Geels, F.W., Kern, F., Fuchs, G., Hinderer, N., Kungl, G., Mylan, J., Neukirch, M. et al (2016) 'The enactment of socio-technical transition pathways: A reformulated typology and a comparative multi-level analysis of the German and UK low-carbon electricity transitions (1990–2014)', *Research Policy*, 45(4): 896–913.

Gen, S. and Wright, A.C. (2013) 'Policy advocacy organizations: A framework linking theory and practice', *Journal of Policy Practice*, 12(3): 163–93.

Gen, S. and Wright, A.C. (2018) 'Strategies of policy advocacy organizations and their theoretical affinities: Evidence from Q-Methodology', *Policy Studies Journal*, 46(2): 298–326.

Hassler, C.M., Mende, M., Scott, M.L. and Bolton, L.E. (2025) 'The prosocial ambassador effect: Adopting an ambassador role increases sustainable behavior', *Journal of Marketing*, 89(1): 19–38.

Hedling, E. and Meeuwisse, A. (2015) 'The European Citizens' Initiative stage: A snapshot of the cast and their Acts', in H. Johansson, and S. Kalm (eds) *EU Civil Society: Patterns of Cooperation, Competition and Conflict*: Springer, pp 210–28.

Henriksen, L.S. and Svedberg, L. (2010) 'Volunteering and social activism: Moving beyond the traditional divide', *Journal of Civil Society*, 6(2): 95–8.

Hoff, J. and Gausset, Q. (2015) 'Community governance and citizen-driven initiatives in climate change mitigation: An introduction', in J. Hoff and Q. Gausset (eds) *Community Governance and Citizen-Driven Initiatives in Climate Change Mitigation*: Routledge, pp 1–6.

Hojnacki, M., Kimball, D.C., Baumgartner, F.R., Berry, J.M. and Leech, B.L. (2012) 'Studying organizational advocacy and influence: Reexamining interest group research', *Annual Review of Political Science*, 15(1): 379–99.

Hooghe, L. and Marks, G. (2001) *Multi-level Governance and European Integration*, Rowman & Littlefield.

Hyde, M.K., Dunn, J., Bax, C. and Chambers, S.K. (2016) 'Episodic Volunteering and Retention', *Nonprofit and Voluntary Sector Quarterly*, 45(1): 45–63.

International Labour Organization (2016) 'Volunteer work', [online] Available from: https://www.ilo.org/resource/volunteer-work [Accessed 14 August 2025].

Kelle, N., Kausmann, C., Schauer, J., Lejeune, C., Wolf, T., Simonson, J. et al (2021) German Survey on Volunteering-Deutscher Freiwilligensurvey (FWS) 2019: Survey Instrument-English Version. German Centre of Gerontology.

Kingdon, J. (1984) *Agendas, Alternatives, and Public Policies*, Longman.

Kivimaa, P., Bergek, A., Matschoss, K., and Van Lente, H. (2020) 'Intermediaries in accelerating transitions: Introduction to the special issue', *Environmental Innovation and Societal Transitions*, 36: 372–7.

REFERENCES

Klaever, A. and Verlinghieri, E. (2025) 'Who is (not) in the room? An epistemic justice perspective on low-carbon transport transitions', *Journal of Environmental Policy & Planning*, 27(2): 79–94.

Knill, C. and Liefferink, D. (2007) *Environmental Politics in the European Union: Policy-Making, Implementation and Patterns of Multi-Level Governance*. Manchester University Press.

Knill, C. and Tosun, J. (2009) 'Hierarchy, networks, or markets: How does the EU shape environmental policy adoptions within and beyond its borders?' *Journal of European Public Policy*, 16(6): 873–94.

Kohler-Koch, B. (2015) 'Participation by invitation: Citizen engagement in the EU', in T. Poguntke, S. Rossteutscher, R. Schmitt-Beck, and S. Zmerli (eds) *Citizenship and Democracy in an Era of Crisis: Essays in Honour of Jan W. van Deth*: Routledge, pp 206–23.

Kostadinova, P. (2015) 'Improving the transparency and accountability of EU institutions: The impact of the Office of the European Ombudsman', *Journal of Common Market Studies*, 53(5): 1077–93.

Krämer, L. (2009) 'The environmental complaint in EU law', *Journal for European Environmental & Planning Law*, 6(1): 13–35.

Kulovesi, K., Oberthür, S., Van Asselt, H. and Savaresi, A. (2024) 'The European Climate Law: Strengthening EU procedural climate governance?', *Journal of Environmental Law*, 36(1): 23–42.

Lee, Y. and Brudney, J.L. (2012) 'Participation in formal and informal volunteering: Implications for volunteer recruitment', *Nonprofit Management and Leadership*, 23(2): 159–80.

Lenschow, A., Burns, C. and Zito, A. (2020) 'Dismantling, disintegration or continuing stealthy integration in European Union environmental policy?', *Public Administration*, 98(2): 340–8.

Lorenzoni, I., Jordan, A.J., Sullivan-Thomsett, C. and Geese, L. (2025) 'A review of National Citizens' Climate Assemblies: Learning from deliberative events', *Climate Policy* (early view): 1–17.

Machin, A. and Tan, E. (2024) 'Green European citizenship? Rights, duties, virtues, practices and the European Green Deal', *European Politics and Society*, 25(1): 152–67.

Marqués Ruizen, C. (2024) 'Chapter 26: European Climate Pact and Climate Ambassadors', [blog] Available from: https://carmenmarq uesembajadorapactoclimático.eu/blog/chapter-26-european-clim ate-pact-and-climate-ambassadors/ [Accessed 14 August 2025].

McKeever, B.W., McKeever, R., Choi, M. and Huang, S. (2023) 'From advocacy to activism: A multi-dimensional scale of communicative, collective, and combative behaviors', *Journalism & Mass Communication Quarterly*, 100(3): 569–94.

Merrilees, B., Miller, D. and Yakimova, R. (2020) 'Volunteer retention motives and determinants across the volunteer lifecycle', *Journal of Nonprofit & Public Sector Marketing*, 32(1): 25–46.

Meyer, M. (2010) 'The rise of the knowledge broker', *Science Communication*, 32(1): 118–27.

Miller, G.J. (2005) 'The political evolution of principal-agent models', *Annual Review of Political Science*, 8: 203–25.

Nadkarni, N.M., Weber, C.Q., Goldman, S.V., Schatz, D.L., Allen, S. and Menlove, R. (2019) 'Beyond the deficit model: The ambassador approach to public engagement', *BioScience*, 69(4): 305–13.

Nagel, M., Gall, A. and Tosun, J. (2025) 'The "hottest ever January" in Germany: Farmers' protests and the discourse on agriculture and food production', *Politics and Governance*, 13(1): 1–23.

Nesbit, R. (2017) 'Advocacy recruits: Demographic predictors of volunteering for advocacy-related organizations', *Voluntas*, 28(3): 958–87.

Newig, J. and Koontz, T.M. (2014) 'Multi-level governance, policy implementation and participation: the EU's mandated participatory planning approach to implementing environmental policy', *Journal of European Public Policy*, 21(2): 248–67.

Oberthür, S. and Dupont, C. (2021) 'The European Union's international climate leadership: Towards a grand climate strategy?' *Journal of European Public Policy*, 28(7): 1095–114.

Oberthür, S. and Kulovesi, K. (2025) 'Accelerating the EU's climate transformation: The European Green Deal's Fit for 55 Package unpacked', *Review of European, Comparative & International Environmental Law*, 34(1): 7–22.

REFERENCES

Oberthür, S. and Roche Kelly, C. (2008) 'EU leadership in international climate policy: Achievements and challenges', *The International Spectator*, 43(3): 35–50.

Omoto, A.M. and Snyder, M. (2002) 'Considerations of community', *American Behavioral Scientist*, 45(5): 846–67.

Oser, J., Grinson, A., Boulianne, S. and Halperin, E. (2022) 'How political efficacy relates to online and offline political participation: A multilevel meta-analysis', *Political Communication*, 39(5): 607–33.

Otjes, S. and de Jonge, L. (2024) 'The Netherlands: Political developments and data in 2023: Two landslide populist victories', *European Journal of Political Research Political Data Yearbook*, 63(1): 317–36.

Parks, L., Della Porta, D. and Portos, M. (2023) 'Environmental and climate activism and advocacy in the EU', in T. Rayner, K. Szulecki, A.J. Jordan, and S. Oberthür (eds) *Handbook on European Union Climate Change Policy and Politics*: Edward Elgar Publishing, pp 98–112.

Parreira, M.J. and Pires, I. (2025) 'Empowering rural communities on rural pact implementation: A human–ecological perspective on social innovation and rural young entrepreneurship', Proceedings, 113(1): article 2.

Patterson, J.J. (2023) 'Backlash to climate policy', *Global Environmental Politics*, 23(1): 68–90.

Petridou, E. and Mintrom, M. (2021) 'A research agenda for the study of policy entrepreneurs', *Policy Studies Journal*, 49(4): 943–67.

Porto de Oliveira, O. (2020) 'Policy ambassadors: human agency in the transnationalization of Brazilian social policies', *Policy & Society*, 39(1): 53–69.

Principi, A., Schippers, J., Naegele, G., Di Rosa, M. and Lamura, G. (2016) 'Understanding the link between older volunteers' resources and motivation to volunteer', *Educational Gerontology*, 42(2): 144–58.

Quittkat, C. and Finke, B. (2008) 'The EU Commission consultation regime', in B. Kohler-Koch, D. de Bièvre, and W. Maloney (eds) *Opening EU-Governance to Civil Society*, pp 183–222. Mannheim Centre for European Social Research (MZES).

Reas, J., Leung, Y.-F. and Cajiao, D. (2023) 'Ambassadors, stewards, advocates – Is engagement of polar tourists in conservation symbolic or substantive? A scoping review', *Frontiers in Sustainable Tourism*, 2: 1263644.

Rehmet, J. and Dinnie, K. (2013) 'Citizen brand ambassadors: Motivations and perceived effects', *Journal of Destination Marketing & Management*, 2(1): 31–8.

Rzepka, D.S. and Tosun, J. (2025) *European Climate Pact Ambassador Survey (ECPAS): Dataset*, heidata, [online] Available from: https:// doi.org / 10.11588/ DATA/ ZIDN7D [Accessed 8 July 2025].

Sandmann, L., Bülbül, E., Castano-Rosa, R., Hanke, F., Großmann, K., Guyet, R. et al (2024) 'The European Green Deal and its translation into action: Multilevel governance perspectives on just transition', *Energy Research & Social Science*, 115: 103659.

Schulze, K. and Tosun, J. (2013) 'External dimensions of European environmental policy: An analysis of environmental treaty ratification by third states', *European Journal of Political Research*, 52(5): 581–607.

Shore, J. and Tosun, J. (2019) 'Personally affected, politically disaffected? How experiences with public employment services impact young people's political efficacy', *Social Policy & Administration*, 53(7): 958–73.

Snyder, M. and Omoto, A.M. (2008) 'Volunteerism: Social issues perspectives and social policy implications', *Social Issues and Policy Review*, 2(1): 1–36.

Søgaard Jørgensen, M. and Pedersen, S.R. (2015) 'Climate ambassador programmes in municipalities', in J. Hoff, and Q. Gausset (eds) *Community Governance and Citizen-Driven Initiatives in Climate Change Mitigation*: Routledge, pp 107–29.

Sommermann, K.P. (2015) 'Citizen participation in multi-level democracies: An Introduction', *Citizen Participation in Multi-level Democracies*: Brill | Nijhoff, pp 1–12.

Stadelmann-Steffen, I. and Gundelach, B. (2015) 'Individual socialization or polito-cultural context? The cultural roots of volunteering in Switzerland', *Acta Politica*, 50(1): 20–44.

REFERENCES

Tarrow, S. (1988) 'National politics and collective action: Recent theory and research in Western Europe and the United States', *Annual Review of Sociology*, 14(1): 421–40.

Tatham, M. and Peters, Y. (2023) 'Fueling opposition? Yellow vests, urban elites, and fuel taxation', *Journal of European Public Policy*, 30(3): 574–98.

Tobin, P., Torney, D. and Biedenkopf, K. (2023a) 'EU climate leadership: domestic and global dimensions', in T. Rayner, K. Szulecki, A.J. Jordan, and S. Oberthür (eds) *Handbook on European Union Climate Change Policy and Politics*: Edward Elgar Publishing, pp 187–200.

Tobin, P., Farstad, F.M. and Tosun, J. (2023b) 'Intermediating climate change: the evolving strategies, interactions and impacts of neglected "climate intermediaries"', *Policy Studies*, 44(5): 555–71.

Tosun, J. (2022a) 'Addressing climate change through climate action', *Climate Action*, 1(1): 1–8.

Tosun, J. (2022b) 'European Citizens' Initiative', in P.R. Graziano, and J. Tosun (eds) *Elgar Encyclopedia of European Union Public Policy*: Edward Elgar Publishing, pp 130–37.

Tosun, J. (2022c) 'What role for climate pact ambassadors? A policy process perspective', *European View*, 21(2): 171–7.

Tosun, J., Béland, D. and Papadopoulos, Y. (2022) 'The impact of direct democracy on policy change: insights from European citizens' initiatives', *Policy & Politics*, 50(3): 323–40.

Tosun, J. (2023) 'The European Union's climate and environmental policy in times of geopolitical crisis', *Journal of Common Market Studies*, 61(S1): 147–56.

Tosun, J., Geese, L. and Lorenzoni, I. (2023a) 'For young and future generations? Insights from the web profiles of European Climate Pact Ambassadors', *European Journal of Risk Regulation*, 4(4): 747–59.

Tosun, J., Pollex, J. and Crumbie, L. (2023b) 'European climate pact citizen volunteers: strategies for deepening engagement and impact', *Policy Design and Practice*, 6(3): 344–56.

Tosun, J., Tobin, P. and Farstad, F.M. (2023c) 'Intermediating climate change: conclusions and new research directions', *Policy Studies*, 44(5): 687–701.

Vecchione, M. and Caprara, G.V. (2009) 'Personality determinants of political participation: The contribution of traits and self-efficacy beliefs', *Personality and Individual Differences*, 46(4): 487–92.

Von Homeyer, I., Oberthür, S. and Dupont, C. (2022) 'Implementing the European Green Deal during the evolving energy crisis', *Journal of Common Market Studies*, 60(S1): 125–36.

Wagner, P.M., Ocelík, P., Gronow, A., Ylä-Anttila, T. and Metz, F. (2023) 'Challenging the insider outsider approach to advocacy: How collaboration networks and belief similarities shape strategy choices', *Policy & Politics*, 51(1): 47–70.

Weller, M.-P. and Tran, M.-L. (2022) 'Climate litigation against companies', *Climate Action*, 1(1).

Williams, C.C. (2008) 'Developing a culture of volunteering: Beyond the third sector approach', *Journal of Voluntary Sector Research*, 1(1): 25–44.

Wilson, J. (2000) 'Volunteering', *Annual Review of Sociology*, 26(1): 215–40.

Wynne, B. (2007) 'Public participation in science and technology: Performing and obscuring a political–conceptual category mistake', *East Asian Science, Technology and Society: An International Journal*, 1(1): 99–110.

Ziemek, S. (2006) 'Economic analysis of volunteers' motivations – A cross-country study', *The Journal of Socio-Economics*, 35(3): 532–55.

Zito, A.R., Burns, C. and Lenschow, A. (2019) 'Is the trajectory of European Union environmental policy less certain?', *Environmental Politics*, 28(2): 187–207.

Index

References to figures appear in *italic* type. References to endnotes show both the page number and the note number (150n6).

A

activism 14, 15, 48, 51, 52, 55–7, 69, 82, 102, 139
  contentious forms of 15, 32, 145
  environmental activism 14, 55, 86, 145, 148
  online activism 117, 130, 138
  see also advocacy; climate action; volunteering
advocacy 14–15, 43, 54–8, 62–5
  direct advocacy strategies 56–7, 62, 69, 81–3
  individual versus organizational 58, 62
  indirect advocacy strategies 56–7, 69, 81–3
  insider versus outsider strategies 63–4
  policy advocacy 43, 104–20, *106–7*, 137–41
  see also activism; strategies
age differences, and motivation 39, 59, 79, 99–102, *100–1*, 130, 137
agents of change 42, 55
Alemanno, A. 33
Alliance for Sustainability Leadership in Education 150n1(Ch 2)
altruistic motivation 52, 53, 96, 122, 131–2, 134–5
  by gender 59–60, 102
  moral obligation 61–2, 91, 102, 134
  by professional background 97–9, 102
  societal need 91, 99, 135
Australian Bureau of Statistics 81, 99

B

behavioural change 20, 29, 31, 42
Beierlein, C. 83
Belgium 88
Beyers, J. 82, 140, 141, 145
Binderkrantz, A. 56, 69
Boasson, E.L. 3
brand ambassadors 23
Bulgaria 105
Burns, C. 150n1(Ch1)

C

capacity building 9, 25, 29, 43, 50
career development, as motivation 53, 59, 94–5, 103, 136
cause ambassadors 22–3
citizen engagement 6–7, 24, 40–4, 144
  barriers to 13
  bottom-up 12, 40
  deliberative democracy 38
  in European Union 32–9
  invited versus claimed spaces 11–12, 33
  legitimacy of 40
  top-down 12
Civil Dialogue 33–4
civil society 7, 14, 30, 33–4, 44, 47, 60, 63, 105, 108, 120, 139
Clean Industrial Deal (CID) 149
climate action 6, 9–10, 15, 23, 31–2, 42, 55, 94, *109–13*
  local level 12–13, 42, 129
  multilevel 129, 145–6
  see also activism

Climate Action Tracker 38, 150n6
climate change
　awareness of 8, 27–9, 31, 73, 93
　politicization of 4–5
　urgency of 31, 91, 99, 135
climate governance 7, 9, 11, 19, 20–1, 24, 37, 44, 54, 132, 145, 148
climate intermediaries 43, 45, 147–8
climate policy 3–6, 20, 42, 52, 148
Climate Pact Ambassadors (CPAs) 6–7, 10–17, 22–32, 49–54, 72–9, 88–91
　activities of 27–9, 31–2, 104–5
　benefits of 27, 29, 42–3
　eligibility criteria 25–6, 29–31, 44, 77
　experiences of *see* experience of CPAs
　mandate duration 25, 128
　mandate types (individual vs organizational) 61–2, 102, 105, 108, 114–16
　motivations of *see* motivations of CPAs
　population and sample 88–91, *89–90*
　professional backgrounds of *see* professional backgrounds
　public profiles of 72–6
　public visibility 13, 73
　representative function 31–2
　strategies of *see* strategies of CPAs
　volunteering *see* volunteering
　working areas 27–9, *28*
communication 14–15, 27, 30, 31, 34–5, 51, 117, 138
community 6, 12–13, 40–2, 103, 129, 145
Compact of Mayors 152n12
Conference on the Future of Europe 38
consultations, public 34–5

contacting politicians 114, 117, 138, 140
courts 114, 117, 120, 130, 140
Covenant of Mayors for Climate and Energy 129, 152n12

D

deliberative democracy 38
Della Porta, D. 12
democracy 5, 9, 20, 32–6, 35, 38, 41, 144
demonstrations 14, 55, 57, 82, 105, 108, 114, 120, 139
Denmark 23, 104
Department for Education 150n1(Ch 2)
digital infrastructure 39
digital strategies *see* online activism; social media

E

egoistic motivation 59–60, 91, 93–6, 122, 132
　emotional reward 93, 96, 103, 125, 135–6
　interest in activity 94, 96, 103, 136
　personal satisfaction 52, 93, 95, 96
eligibility criteria, for CPAs 25–6, 29–31, 44, 77
emissions trading systems 3
emotional reward 93, 96, 103, 125, 135–6
environmental activism 14, 55, 145
environmental degradation 1, 5, 8, 20, 29, 31, 46, 73, 93, 132
environmental policy 6, 24, 32, 52, 55
environmentalism 14, 55
European Citizens' Initiative (ECI) 7, 36–8, 43
European Citizens' Panels 38–9
European Climate Law 9

INDEX

European Climate Pact (ECP) 5–8, 148
 and citizen engagement 32, 39–44
 and just transition 29
 benefits of 42–3
 components of 8–11
 effectiveness of 20–1, 46
 features of 24–32
 Friends of the Pact 11
 governance tool 24, 46
 institutional design 9, 12
 objectives 8–9, 15
 online platform 27, 71, 73–4, 76, 85
 operational scope 41
 Partners of the Pact 10–11
 satellite events 10
 Secretariat 25, 27, 43, 77, 142, 144
European Climate Pact Ambassador Survey 72, 76–9, 88, 132
European Commission 2, 5, 12, 15, 18–19, 23–32, 42, 44–5, 128, 137–8, 142–5, 150
European Economic and Social Committee 33
European Green Deal (EGD) 2–6, 8, 12, 15, 20, 23–4, 29, 31, 41–2, 44–6, 54, 56, 132, 148–50
European Ombudsman 33–4
European Parliament 7, 37, 105
European Union (EU)
 Climate Action Academy 29, 142
 multilevel governance 1, 12–13, 129, 145–6
 policy approach 1–4
experiences of CPAs 68, 121–9, *121*, *126*
 and motivation 65, 122–5, *123–4*
 negative experiences 121–2, 127, 141
 and political efficacy 65–7, 125–7
 positive experiences 121–2, 125, 141
 suggestions for improvement 127–9, 142–3
external political efficacy 65, 66–7, 83–4, 125–7, 141–2

F

farmers' protests 4–5
Felicetti, A. 12
financial compensation, absence of 16, 27, 49
formal volunteering 50–1
France 4, 35
Fridays for Future 37–8
Friends of the Pact 11
Futurium 39

G

Gausset, Q. 24
gender differences, and motivation 59–60, 79, *100–1*, 102–3, 130, 137–8
Germany 3, 88
Global Ambassadors 23
Global Covenant of Mayors for Climate and Energy 152n12
grassroots mobilization 6, 42, 55
Greece 104–5
green European citizenship 6, 132, 134, 135, 137
green deal diplomacy 23
greenhouse gas (GHG) emissions 2, 38, 46, 73

H

Hassler, C.M. 151n1(Ch 4)
Henriksen, L.S. 47, 54, 55
Hoff, J. 24
Hoekstra, Wopke 25, 105

I

influencers 26, 132
informal volunteering 50–1

insider strategies 14, 63–4, 67
institutional backing 24, 62
institutional transparency 35–6
intermediation 148
internal political efficacy 66, 83, 125–7, 141–2
investment motivation 52–3, 91, 96, 122, 130–1, 142
  career advancement 59, 94–5, 103, 136
  networking 29, 43, 51, 94–5, 103, 127, 136, 142
  skill development 95, 125, 136
invited spaces 11–12, 32, 33
Ireland 88
Italy 88, 104

J

just transition 5–6, 29, 73, 105

K

Kingdon, J. 55
Klimafolkemødet (The Climate People's Meeting) 104
knowledge brokers 51
knowledge sharing 9, 14, 26, 32, 73
Kohler-Koch, B. 33
Koontz, T.M. 53

L

legitimacy 5, 20, 40, 41
Lisbon Treaty 35–6
lobbying 19, 56, 63–4

M

Maastricht Treaty 33
Machin, A. 6, 134, 136, 138
Malta 88
Marqués Ruizen, C. 132, 141, 142
Members of Parliament (MPs) 114–19, 120, 130–1, 138, 140

ministers 114–19, 120, 130–1, 140
Mintrom, M. 55
mission statements 73–4, 134
moral obligation 61–2, 91, 102, 122, 134
motivations of CPAs 51–2, 67, 91–104, *92*, 122–5, 133–7
  by age 59, 99–102, 130, 137
  altruistic 59–60, 91, 95–6, 102, 122, 131–2, 134–5
  egoistic 59–60, 93–6, 103, 122, 125, 132, 135–6
  by gender 59–60, 99–102, 130, 137–8
  investment 52–3, 59, 94–5, 96, 103, 122, 125, 130–1, 136, 142
  by mandate type 61–2, 99–102, *100–1*, 130, 137
  measurement of 80–1
  by professional background 60–1, 97–9, 103–4, 130–1, 137
  tripartite framework 52–3, 68, 133
multilevel political system 12–13, 30, 129, 145

N

national coordinators 25, 26, 129
net-zero objective 2, 8, 38, 73
Netherlands, the 35, 104
networking 29, 43, 51, 94–5, 103, 127, 136, 142
Newig, J. 53
non-governmental organizations (NGOs) 14, 30, 60, 63, 108, 117, 120, 138, 139, 140
non-profit sector 60, 88, 96, 108

O

Omoto, A.M. 54, 70, 147, 148
online activism 114, 117, 138, 139
online platform
  ECP website 27, 71, 73–4, 76, 85
  public profiles on 72–6, 93, 134

INDEX

operationalization 49, 52, 67, 79–80, 86
  of experience 83–4
  of motivation 80–1
  of strategies 81–3
opinion leaders 26
outsider strategies 14, 63, 67

P

Paris Agreement 23, 37–8
Parks, L. 14, 15, 48, 139
Partners of the Pact 10–11
Pedersen, S.R. 7
Peer Parliaments 32, 105, 151–2n7
petitions 33, 114, 117, 138, 139
  right to petition 33
  signing 114, 138, 139, 140
Petridou, E. 55
policy advocacy *see* advocacy
policy ambassadors 22
policy entrepreneurs 55
policy makers 3–4, 15, 42, 56, 128
  access to 15, 42, 63, 114–16, 129, 140–1
  responsiveness of 66
political efficacy 65–7, 70
  external 66–7, 83–4, 125–7, 141–2
  internal 66, 83, 125–7, 141–2
political parties, working in 108, 114, 120, 138, 139
political stakeholders, access to
  courts 114, 117, 120, 131, 140
  by mandate type 114–16, *115–16*, 131
  ministers and MPs 114, 117, 120, 130–1, 140
  by professional background 117–20, 118–19, 140
politicization of climate policy 4–5
Porto de Oliveira, O. 22
principal–agent model 61
Principi, A. 60
private consumption model 52
private sector 60, 63–4, 88, 96, 108, 117, 139, 140

professional background 60–5, 79, 88–91, 96–9, *97–8*, 103–4, 108–14, 117–20, 130–1, 137–40
  academia/education 60, 88, 96, 108, 117, 139
  civil society 60, 63, 108, 117, 139
  non-profit sector 60, 88, 96, 108
  private sector 60, 63–4, 88, 96, 108, 117, 139, 140
  public sector 60, 63, 88, 96, 108, 114, 117, 120, 140
  students 63, 88, 96, 103, 139
protests 4–5, 32, 55, 105, 139
public office holders 26
public policies 3–4, 20
public profiles 72–6, 93, 134
public sector 60, 63, 88, 96, 108, 114, 117, 120, 140
public support 5

R

Race to Zero campaign 23
Reas, J. 138
recruitment 25–6, 44, 147
Regulation (EC) No 1049/2001 35
renewal of mandate 25, 40, 128
representative function 31–2
retention 53–4, 128
Rural Pact 8

S

satisfaction, in volunteering 52, 54, 93, 95–6, 122, 141–2
selection process 12, 24, 136, 137
Shore, J. 66
skill development 95, 125, 136, 142
Slovak Republic 105
Snyder, M. 54, 70, 147, 148
social acceptance 4, 20
social media 27, 75, 117, 131, 138, 139
social movements 14, 37, 55
social network analysis 146, 147
social principles 91, 102, 131, 133–4, 137, 138

societal need 91, 99, 102, 135, 137
Søgaard Jørgensen, M. 7
Spain 88, 105
'Stop TTIP' Initiative 36
strategies of CPAs 62–5, 67–8, 81–3, 104–20, 137–40
  direct advocacy 56–7, 62, 69, 81–3
  indirect advocacy 56–7, 69, 81–3
  insider versus outsider 63–4, 67
  by mandate type 105, 108, 114–16, 131, 138–9
  measurement of 81–3
  by professional background 62–5, 108–14, 117–20, 131, 139–40
students 63, 88, 96, 103, 139
Svedberg, L. 47, 54, 55
supranational organization 18–19, 132
sustainability 2, 22–3, 31, 39, 41, 44, 148
Sweden 105

T

Tan, E. 6, 134, 136, 138
Tatham, M. 3
Technical Expert Group 129
Timmermans, Frans 25
Tobin, P. 150n1(Ch 1)
Tosun, J. 27–9, 30, 60, 61, 66, 76, 77, 78, 79, 102, 128, 134, 135, 137
training 26, 29, 43, 50, 127, 136, 142
transformative change 3, 54–5

U

United Kingdom (UK) 3, 23
United Nations
  Act Now initiative 9
  Framework Convention on Climate Change (UNFCCC) 23, 38, 129
University of Reading 150n1(Ch 2)

V

Vienna Convention on Diplomatic Relations 22
volunteer process model 54, 65, 70, 147, 148
volunteering 47–54, 68, 80, 91, 129, 133, 145
  and advocacy 54–8
  definition 49, 80
  episodic versus ongoing 50
  formal versus informal 50–1
  motivation for 50–4, 80–1, 91–104
  research on 47, 51, 60, 145
  retention of 53–4
von der Leyen, Ursula 1–2, 25

W

Wagner, P.M. 57, 64
Wilson, J. 49

Y

yellow vests (*gilets jaunes*) movement 4
Your Voice in Europe 39

Z

Ziemek, S. 52–3, 59, 65, 68, 80, 81, 91, 96, 99, 133, 134, 138

www.ingramcontent.com/pod-product-compliance
Lightning Source LLC
Chambersburg PA
CBHW051549020426
42333CB00016B/2174